T0344453

Cleaning Validation

Pharmaceutical manufacturers and upper management are encouraged to meet the challenges of the science-based and risk-based approaches to cleaning validation. Using some of the principles and practices in this volume will help in designing a more effective and efficient cleaning validation program.

Features
- Timely coverage of cleaning validation for the pharmaceutical industry, a dynamic area in terms of health-based limits.
- The author encourages pharmaceutical manufacturers, and particularly upper management, to meet the challenges of the science-based and risk-based approaches to cleaning validation.
- Draws on the author's vast experience in the field of cleaning validation and hazardous materials.
- Discusses EMA vs. ISPE on Cleaning Limits and revised Risk-MaPP for highly hazardous products in shared facilities.
- A diverse list of topics from protocol limits for yeasts and molds to cleaning validation for homeopathic drug products.

Cleaning Validation
Practical Compliance Approaches for Pharmaceutical Manufacturing

Destin A. LeBlanc

Consultant

Cleaning Validation Technologies

CRC Press
Taylor & Francis Group
Boca Raton London New York

CRC Press is an imprint of the
Taylor & Francis Group, an **informa** business

First edition published 2023
by CRC Press
6000 Broken Sound Parkway NW, Suite 300, Boca Raton, FL 33487-2742

and by CRC Press
4 Park Square, Milton Park, Abingdon, Oxon, OX14 4RN

ISBN: 978-1-032-43083-6 (hbk)
ISBN: 978-1-032-43173-4 (pbk)
ISBN: 978-1-003-36600-3 (ebk)

DOI: 10.1201/9781003366003

Typeset in Times
by SPi Technologies India Pvt Ltd (Straive)

Contents

Section I   Terminology

Section II   Health-Based Limits

Section III   Limits: General

Section IV   Visually Clean

Section V   Analytical and Sampling Methods

Section VI   Product Grouping

Section VII   Protocols and Procedures

Section VIII   API Manufacture

Section IX   Miscellaneous Topics

Preface

This Volume 5 complements my earlier four books on the same subject. What are presented in this book are modifications and updates of my monthly *Cleaning Memos* originally published on my website, www.cleaningvalidation.com, in the period from January 2017 through December 2020. Each *Cleaning Memo* is presented as a chapter, with the chapters then *organized by common topics*. For example, topics related to setting limits are in one section, those related to sampling in another section, and so forth. The updates made are sometimes relatively simple, and sometimes more detailed. In all cases, I have tried to focus on changes for improving clarity and applicability, as well as to modify the text with new information. There is one appendix with a list of acronyms used in this volume, as well as a second appendix dealing with my shorthand method of expressing limits, just in case you get confused about what I mean by L0, L1, L2, L3, and L4.

A "hot topic" for these four years has continued to be limits, specifically health-based limits. About half of these chapters deal with setting limits in one way or another, so the use of health-based limits will require balanced reading (and thinking) for an overall understanding.

I would also like to encourage pharmaceutical manufacturers, and particularly upper management, to meet the challenges of the science-based and risk-based approaches to cleaning validation (as opposed to a "cook book" approach). Using some of the principles and practices in this volume may help in designing a more effective and efficient cleaning validation program.

I should add a caveat here, much like the caveat before each of the FDA's guidance documents – nothing in this book should be considered mandatory or binding. I have tried to present alternatives where possible. However, achieving the same objectives by utilizing scientifically justified procedures that are applicable to a manufacturer's specific situation is strongly encouraged. In a manufacturing environment where process efficiencies as well as good compliance are mandatory, following "cookie cutter" recipes should be avoided.

Before you *connect* the dots,

make sure you *collect* the dots!

Destin A. LeBlanc
Knoxville, TN

About the Author

Destin A. LeBlanc is a consultant at Cleaning Validation Technologies. He has extensive experience in product development and technical services for cleaning and antimicrobial applications. He is an international lecturer on contamination control and has written widely on cleaning validation topics including four volumes in the "Cleaning Validation: Practical Compliance Solutions for Pharmaceutical Manufacturing" series published by PDA and DHI. He is a member of PDA and ISPE and has trained FDA personnel on cleaning validation. He is a graduate of the University of Michigan and the University of Iowa.

Section I

Terminology

The following six chapters cover issues related to how terms are sometimes defined and used differently. It should be emphasized that while consistent use of terms/definitions within the industry would be helpful, perhaps the more critical issue for individual companies is to use terms/definitions consistently *within* that company and within company documents.

1 Use of the Term "Product"

I have become more concerned in the last few years about the correct and consistent use of words, in particular in cleaning validation documents. This first chapter considers the use of the word "product". What does this refer to? Is this a "generic" use of the word "product", in that it could refer to almost anything? For example, I might refer to it meaning that the "product" could be the active, the "drug product" itself, a starting material, and excipient. If used in this generic sense, what constitutes a "product" may be left up to the interpretation of the reader/auditor. If what I really mean is the "drug product", then that term ("drug product") should really be used. If I really mean product in a generic sense, then there may be other words to use so that I avoid any possible misinterpretation that I am referring to the "drug product". Examples of other words that may fall under the generic umbrella of "product" might include "residue" and "materials".

An example of this confusing use of the word "product" is as old as limit setting in cleaning validation. The classic "Fourman and Mullen" 1993 paper on limits is a reference to a default limit as "No more than 10 ppm of any product will appear in another product" [Fourman and Mullen, 1993]. I have seen this interpreted as "10 ppm of one drug product in another drug product". Clearly, this was not the intent of the authors. What was intended was "No more than 10 ppm of any active ingredient will appear in another drug product". At least the latter is how this has been historically applied.

One way to prevent confusion is to have a glossary of typical terms that are used. For example, in a "drug product" manufacturing facility, I might have both "drug product" and "product" defined. A definition of "drug product" might be the one typically used [FDA, 2020]:

I. Drug product: A finished dosage form that contains an active pharmaceutical ingredient, generally, but not necessarily, in association with inactive ingredients.

Then a definition of "product" might be something like:

II. Product: As referred to in this document, "product" is a generic term referring to any drug product, drug product intermediate, or active ingredient.

On the other hand, in a "drug substance" (Active Pharmaceutical Ingredient or API) manufacturing facility, I might have definitions for "drug substance", but the following might be one definition for "Product":

DOI: 10.1201/9781003366003-2

III. Product: As referred to in this document, "product" is a generic term referring to any drug substance, crude drug substance, reactant, intermediate, or
by-product.

Having such a definition in cleaning validation documents for drug substance
manufacturing facilities is important so that auditors don't assume that the
cleaning validation approach for a drug product applies to the facility.

Realize that there may be situations where even the generic use of the term
"product" might not be appropriate. For example, in determining whether a surface is visually clean, just saying the surface is free of "any product" might not be
good enough, unless the concept of product were expanded to include cleaning
agents. There are two possible approaches to address this. Here is one option for
drug product manufacturing where the term "product" is not used at all:

IV. Visually clean: Free from visible residues including drug products, active
ingredients, drug product intermediates, excipients, cleaning agents, foreign materials, and by-products of the cleaning process under defined viewing conditions.

If I had my previous definition (II) of "product", then this could be shortened to:

V. Visually clean: Free from visible residues including products, excipients,
cleaning agents, foreign materials, and by-products of the cleaning process
under defined viewing conditions.

In definitions IV and V, I do realize the term "by-product" is used, which means I
should also have a suitable definition for "by-product".

Here is an option for "a definition for visually clean" for drug substance
manufacture:

VI. Visually clean: Free from visible residues including drug substances, reactants, intermediates, by-products, cleaning agents, and foreign materials
under defined viewing conditions.

Note that in this case, by-products could include by-products of a manufacturing
step as well as by-products of a cleaning step.

As long as we are considering these definitions, it might also help to mention
the difference between an "active ingredient" and an "active pharmaceutical
ingredient". As is typically used in the industry, an "active ingredient" is an API
in a drug product (in the formulated drug product) [FDA, 2020], while an "active
pharmaceutical ingredient" is the drug substance apart from (or before it is formulated into) a drug product [FDA, 2020]. To the layperson, this sounds like both
terms are the same. Even to those in the drug industry, these terms may be used
loosely as we talk to each other. However, as given in our formal cleaning validation documents (such as cleaning validation master plans, SOPs, protocols, and

risk assessments) it is definitely better to use the terms correctly as defined in applicable glossaries.

Here is one more related clarification. Sometimes we use the word "drug" quite apart from specifying whether it is a "drug product" or a "drug substance". Here is the definition of "drug" in the Federal Food Drug and Cosmetic Act [FD&C Act, 2021]:

> The term "drug" means (A) articles recognized in the official United States Pharmacopeia, official Homeopathic Pharmacopeia of the United States, or official National Formulary, or any supplement to any of them; and (B) articles intended for use in the diagnosis, cure, mitigation, treatment, or prevention of disease in man or other animals; and (C) articles (other than food) intended to affect the structure or any function of the body of man or other animals; and (D) articles intended for use as a component of any articles specified in clause (A), (B), or (C).

After all this, here is a final concern on glossaries. Some companies like to have glossaries in each document, and some companies like to have one corporate glossary that all groups within the company refer to. The advantage of the first option is that terms are readily accessible as the document is read. A possible downside is that the same term may be defined differently by different groups within a company. The advantage of the latter approach is that consistency is assured across the firm. A possible downside is that a term one group might want to use might already be defined in an inconsistent manner in the corporate glossary.

After all this, I must confess that my writings have not always followed what I am saying here is good practice. So you are right to say, "Physician, heal thyself".

The above chapter is based on a *Cleaning Memo* originally published in May 2019.

REFERENCES

FD&C Act, USA Federal Food Drug and Cosmetic Act 21 US code 321 Section 201(g) (1). January 5, 2021 (https://www.law.cornell.edu/uscode/text/21/321, accessed May 3, 2021).

FDA, Code of Federal Regulations, Ttile 21, Part 210, Section 210.3 (b)(4), Revised April 1, 2020 (https://www.accessdata.fda.gov/scripts/cdrh/cfdocs/cfcfr/CFRSearch.cfm?fr=210.3#:~:text=(4)%20Drug%20product%20means%20a,in%20association%20with%20inactive%20ingredients, accessed May 3, 2021).

Fourman, G.L.; Mullen, M.V. "Determining Cleaning Validation Acceptance Limits for Pharmaceutical Manufacturing Operations". *Pharm. Technol.* 1993, 17, pp. 54–60.

2 Use of the Terms Grouping and Matrixing

The terms "grouping" and "matrixing" are currently used to refer to the same concept of using a "worst-case" product in a cleaning validation protocol to represent a selection of other products where no or limited protocol runs may be performed for those other products. Successful validation of the cleaning process for that worst-case product constitutes or covers validation of cleaning of those other products. There are generally certain "rules" to follow if this approach is used. However, the point of this chapter is not to go into those rules [LeBlanc, 2006; also Chapters 33 and 34 in this volume]. Rather the purpose is to describe how in the early days of cleaning validation, and even now to a great extent, the term "matrixing" was used to describe something significantly different in terms of this approach for products.

In the early days of cleaning validation (the 1990s), limits were set based on a "matrix" approach [Mendenhall, 1989]. Remember that limits are always set based on the characteristics of the cleaned product (namely toxicity, originally represented by 0.001 of the daily dose of the active, but now HBELs are also used), characteristics of the subsequently manufactured product (namely daily dose of the drug product and batch size), characteristics shared by both products (namely shared equipment surface area), and sampling parameters (for swab sampling this is the swabbed area and the solvent extraction amount) [Fourman and Mullen, 1993]. The following discussion is based on limit values expressed as the concentration in the swab extraction solution, what I typically call an L4b value using my shorthand methods of expressing limits (see Appendix B of this volume for this scheme). Of course, if you want to only express limits as the MAC or MACO (Maximum Allowable Carryover, or what I call L2), then the shared surface area and sampling parameters would not be used.

So what was done in the early days of cleaning validation (and of course is still done now) was called "matrixing". According to the Cambridge English Dictionary, a matrix is "a group of numbers or other symbols arranged in a rectangle that can be used together as a single unit to solve particular mathematical problems" [Cambridge, 2021]. Another definition is "a rectangular display of features characterizing a set of linguistic items, especially phonemes, usually presented as a set of columns of plus or minus signs specifying the presence or absence of each feature for each item" [WordReference, 2021]. The key seems to be the idea of a *rectangular* presentation or analysis.

What was done in the early days of cleaning validation was to calculate the limit of each product based on the possibility of every other product being the

DOI: 10.1201/9781003366003-3

TABLE 2.1
Matrix of Product by Equipment Train

	P	Q	R
A	+	−	−
B	−	+	−
C	+	−	−
D	−	−	+
E	+	−	−
F	+	−	−
G	+	−	−
H	−	+	−
I	−	+	−
J	−	−	+

next product made in the same equipment. This involved the consideration of *two* matrixes. In the examples given, I will consider a simple case of ten products being made on one of three possible equipment trains. The products will be called *A* through *J*, and equipment trains will be called P, Q, and R. A first matrix is presented in Table 2.1, with equipment trains horizontally across the top as columns and products vertically down the side as rows. A plus symbol (+) indicates that the product is made in that equipment train, and a negative symbol (−) indicates the product is *not* made in that equipment. What follows is a discussion of the impact of that Table 2.1 matrix.

This matrix makes clear what products are made in the same equipment, and therefore what products should be considered for subsequent matrixes as limits are calculated. From Table 2.1, it can be clearly seen that Products *A*, *C*, *E*, *F*, and *G* are all made on equipment train P, so that the limits for each of those products should be calculated using each of the other products for that train as the next product.

So for my next matrix, I list the products made in the same equipment train horizontally along the top (these represent the subsequently manufactured product) and also vertically along the side (these represent the cleaned product). Within each intersecting cell, I enter the calculated limit (which could be an L2, L3, or L4 value, depending on my approach). Note that for this example in the Table 2.2

TABLE 2.2
Matrix of Limits among Products on Train P (L4b as ppm)

	A	C	E	F	G
A	N.A.	1.3	1.	0.7	2.3
C	1.7	N.A.	2.0	1.3	3.1
E	1.1	1.4	N.A.	0.6	2,1
F	0.9	0.6	0.8	N.A.	1.5
G	1.5	1.1	1.6	0.8	N.A.

matrix, I will *not* consider the possibility that the same product will follow itself (for example, *A* followed by *A*) but that could be done.

Table 2.2 is the resulting matrix for equipment train P, *using limits as L4b (ppm in the extracted swab sample)*. I could also set up a similar matrix for products made in Q and for products made in R.

Using this matrix for *P*, I can readily see that the lowest limit (for the active of *A*) for cleaning of *A* with *C*, *E*, *F*, and *G* as the next products is 0.7 ppm. Therefore, if I do a validation protocol for *A*, I select that lowest limit considering each of the possible next products for my L4b limit for *A*. I can then select the applicable limit in the same way for each of *C*, *E*, *F*, and *G*.

Okay, so much for how we generally referred to "matrixing" in the early days of cleaning validation. While that use of the term is still applicable today, there is another use of the term "matrixing" that is also used today. This newer use of the term "matrixing" is similar to what I call "grouping". It applies specifically to the use of a matrix to determine the worst-case product in that grouping approach (which is one way, but not the only way, to select that worst-case product). In this situation, my matrix involves each product horizontally along the top (these represent the cleaned product) and a variety of relevant parameters that are used to determine the worst-case product vertically along the side, with a last row being an overall worst-case value. Within each cell, a rating for each parameter for each product is made based on a predetermined scale. In this case, I am rating everything on a scale of 1 to 5 except for dyes/colorants (which is a scale of 1 to 2), with the higher number being the worst cases. The "Overall Rating" in the last row is a result of multiplying the ratings for the individual parameters for that product. Using the products made on *P*, Table 2.3 illustrates this type of matrix.

In this manner, I am using a matrix to determine the worst-case product for grouping purposes. In this example, the worst-case product is *A*.

Some clarifications are in order. First, the examples in Tables 2.1, 2.2, and 2.3 are just examples, but hopefully, they illustrate the *concept* of the different matrixes used. Second, even those companies which use the term "matrixing" to cover what is done in the Table 2.3 example will also typically perform (directly or indirectly) the matrixing concepts illustrated in Tables 2.1 and 2.2. Third, some companies for the Table 2.2 matrix where the equipment surface area is exactly the same for all combinations will just determine the ratios of "batch size" to

TABLE 2.3

Matrix of Worst-Case Parameters among Products on *P*

	A	*C*	*E*	*F*	*G*
Solubility of active	5	3	3	1	4
Concentration of API	3	2	2	1	1
Total solids of product	4	1	3	4	2
Presence of dyes/colorants	1	1	2	2	1
Overall Rating	*60*	*6*	*36*	*8*	*8*

"maximum daily dose" for the next products, and use the lowest ratio among the different next products to calculate the lowest limit for a given product. In this case, there may be no matrix clearly used, but the matrix concept is present in that approach. Fourth, some companies use software to do all three types of matrixes, in which case the matrix determinations are implicit in the software results.

When all is said and done, I still prefer the use of the term "grouping" to cover the situation illustrated in Table 2.3. Part of the reason is to separate it from matrixing done as illustrated in Tables 2.1 and 2.2, which are still applicable even if I don't do a product grouping approach.

In any case, the purpose of this Chapter is to encourage companies to carefully define terms such as "matrixing" and "grouping" so they are used consistently within that company. Careful definitions help avoid inconsistencies within a company as well as help avoid misunderstandings of a company's approach by regulators or auditors.

The above chapter is based on a *Cleaning Memo* originally published in July 2019.

REFERENCES

Cambridge English Dictionary, (2021) https://dictionary.cambridge.org/us/dictionary/english/matrix (accessed May 3, 2021).

Fourman, G.L.; Mullen, M.V. "Determining Cleaning Validation Acceptance Limits for Pharmaceutical Manufacturing Operations". *Pharm. Technol.* 1993, 17, pp. 54–60.

LeBlanc, D.A., "Product Grouping Strategies", in *Cleaning Validation: Practical Compliance Solutions for Pharmaceutical Manufacturing, Volume 1.* Parenteral Drug Association, Bethesda, MD, 2006, pp. 167–172.

Mendenhall, D.W. "Cleaning Validation", *Drug Dev. Ind. Pharm.* 1989, 15(13), pp. 2105–2114.

Word Reference Online English Dictionary, (2021) https://www.wordreference.com/definition/m%C3%A1trix (accessed May 3, 2021).

3 Deviations and Nonconformances

This chapter covers another example where clarity in the use of language might help. It deals with the similarities and differences between a "deviation" and a "nonconformance". How these are defined and how they are handled for cleaning validation purposes might vary among different companies, and those differences might be okay as long as the terms are used appropriately and consistently *within* that company.

As I use the term, a "deviation" occurs when a *procedure* is not carried out correctly. For example, if the wrong cleaning agent or the wrong cleaning agent concentration was used for a validated cleaning process, that would constitute a deviation – what should have been done (according to the cleaning process SOP) was *not* done or was done *incorrectly*.

As I use the term, a "nonconformance" is when the result of some *analytical data* (such as chemical, microbiological, or visual test) is outside of a specified limit or acceptable value. For example, in a protocol, my acceptance limit for the rinse sample for TOC is 1.2 ppm. My measured value is 2.7 ppm. That rinse result is considered a nonconformance because it is above my limit value.

Now there are similarities between deviations and nonconformances. Both are things that are "bad" and are things that cause headaches for cleaning validation scientists. I would like to avoid both. Furthermore, a nonconformance in meeting limits may be caused by a deviation in the execution of a cleaning SOP (although there may be situations where I exceed my protocol limit but there is no related cleaning process deviation, perhaps because the cause is the poor design of my cleaning process). I also may have a deviation in the cleaning process during protocol execution that does not result in a nonconforming residue value, because my cleaning process was a robust design with significant overkill designed into the SOP. Note that in this last situation, I might have a deviation in the cleaning process with no accompanying nonconformance of the residue value, perhaps because of a deviation in the analytical or sampling methods which led to lower residue values.

In the discussion above, while it is possible that a deviation (for example, in the cleaning procedure, in the analytical procedure, or in the sampling procedure) may cause a nonconformance in the measured residue value, it is *not* the case that a nonconformance in the measured residue value may cause a deviation. A clarification here is necessary; it may be the case that a nonconformance *caused* me to go back and investigate what was done in the cleaning, sampling, and analytical procedures, and that investigation caused me to *discover a deviation* that I was not

DOI: 10.1201/9781003366003-4

previously aware of. However, it would probably *not* be inappropriate to suggest that the nonconformance *caused* the deviation.

The next question is what to do when there is a deviation and/or a nonconformance. We will only discuss this in the context of a *protocol* (although the principles might apply to other situations). There are several possibilities here. We can have a procedural deviation *without* a nonconformance for residue values, a nonconformance for residue values *without* an identified deviation, or a procedural deviation *with* a nonconformance for residue values. For the first situation, we will want to find out the root cause of the deviation and identify and implement corrective actions (see discussion below) to keep that deviation from happening again in the future. Then we will want to determine the impact of that deviation on the validity of the protocol. Realize that it *might* be the case that the deviation did not affect the validity of the protocol (such as when the cleaning process temperature was slightly below the specified range); with the successful residue results we might accept that protocol run as a valid (and successful) run, or we might consider it "invalid" and just require an additional protocol without deviations. On the other hand, if our conclusion is that the deviation might have helped in meeting the residue limits, then we are more likely to say that it clearly is not a failed run, but rather an invalid run. In that case, we would still take corrective action to make sure that specific deviation did not occur again, and continue protocol runs until we had three successful consecutive runs.

The second situation involves no identified deviation with a nonconformance in residue results. Bad luck! Unless you can identify a lab error using your OOS (Out of Specification) procedure [FDA, 2006], the only reasonable conclusion is that your cleaning process was not robust enough to have consistent passing results (that is, this is a failed run). In this situation, it is probably best to go back to the design phase. This may mean actual changes in targets or ranges of some of the critical process parameters; for the manual cleaning process, it may also mean a better-written SOP with associated better training.

The third situation (both an identified deviation and a nonconformance in residue results) might not be as bad as it sounds. Clearly, this may also be a case of an invalid run. If there is a good probability that the deviation was the *cause* of the nonconformance, then there is a strong possibility that the implementation of appropriate corrective action will result in successful protocol runs in the future.

It is important in all these cases that approved OOS and CAPA (Corrective and Preventive Actions) procedures be followed. Particularly for CAPA, it is important to note that there are actually three actions (not two) that are part of that process [Brosseau, 2018]. They are:

Correction, which is *fixing* what is wrong. For example, if the residue testing data is non-conforming, I will need to correct that problem with the equipment so it can be released for safe manufacture of another product. This *may* involve recleaning again with the same SOP and retesting to confirm acceptable residues. If a procedural deviation occurred, such as exceeding the dirty hold time (DHT) and I have not initiated any residue testing, I may decide just to clean twice with the cleaning SOP and test for residues after that. This latter situation would be considered an invalid protocol run; however, I would want to consider a Corrective Action to make sure I did not exceed the DHT in the future.

Corrective Action is taking steps to help make sure the *same specific deviation* does not occur *in the future*. Using the example of exceeding the DHT, I might institute some program whereby I was alerted as the maximum DHT was approached. This would *not* help me with my "Correction", but it would help me avoid the same problem in the future.

Preventive Action, which is taking steps to help make sure *other* deviations or problems do not occur in the future. Perhaps during my investigation of a deviation, I identified other possible changes (*not* specifically related to the immediate deviation) that could be made to help ensure that my cleaning process stayed in a "state of control". These may be items that were considered as possible root causes, but were rejected as the root cause for that specific deviation; however, I saw an opportunity to improve my cleaning process as part of "continuous improvement". Preventive Actions are not necessarily associated with any deviation, but may be something new I have learned from current technical literature on cleaning processes that could be implemented to improve my process.

While this discussion of deviations and nonconformances reflects one approach to those terms, let me emphasize that other approaches are possible. In any case, there should be an attempt to assure consistency within a given company or facility.

The above chapter is based on a *Cleaning Memo* originally published in June 2019.

REFERENCES

Brosseau, B. "Corrections, Corrective Actions, and Preventive Actions: Effectively Handling Nonconformances in Compliance with the EU MDR", 2018. https://www.medtechintelligence.com/feature_article/corrections-corrective-actions-and-preventive-actions-effectively-handling-nonconformances-in-compliance-with-the-eu-mdr/ (accessed May 3, 2021).

FDA, *Investigating Out-of-Specification (OOS) Test Results for Pharmaceutical Production*. US Government Printing Office, Washington DC, 2006.

4 Clarifying Terms
Blanks vs. Controls

As used for analytical measurements for cleaning validation studies, the two terms "blanks" and "controls" are different [Vitha et al., 2005], but sometimes are used interchangeably. As I generally use the term, a *blank* is a sample for which the measured value is subtracted from a test sample to determine the analyte in the test sample due to the residue removed from the sampled surface. There are two ways a blank may be used. In one case (for example, in a spectrophotometric analysis), the blank is used to set a "zero" set point for instrument output. Once the zero is set, I then place my test sample in the spectrophotometer and measure the absorbance, which I can then attribute to the residue I want to measure. This technique deals with impurities (for example) in the solvent used for the extraction of a swab. Now if I have a double beam spectrophotometer, the blank sample is measured at the same time as the test sample, and the instrument *subtracts out* the blank value.

A second way a blank is used can be illustrated by the use of TOC (Total Organic Carbon) as the analytical technique. In a test sample analyzed by TOC, there are multiple sources of carbon. What I am really interested in measuring is the carbon *removed from the surface.* But, other sources of carbon include the vial, the water, the swab itself (for swab sampling), and the air in the atmosphere where I prepare the sample. In addition to the test sample, I prepare a blank with just the vial, water, and a swab in the same location where I take my test sample. The analytical lab runs both the test sample and the blank on the TOC instrument and then subtracts the two to arrive at a TOC value representing the residue on the sampled surface.

Okay, what then is a *control?* A control is a sample that is analyzed where I am *expecting* to get a certain value or range of values. This control provides information as to the *validity* of my analysis. If the control is not in the expected range, then something is wrong. What is wrong may be a problem with the instrument, with the reagents used in the analysis, or with the sample preparation. An out-of-specification control calls into question the validity of the analytical results obtained for my test samples. Let's illustrate this again using TOC. One control might be running the analytical procedure on just water in a vial. I know based on past results that this control should be in the range of 30–60 ppb TOC (for example). Another control might be running a KHP (potassium hydrogen phthalate) solution equivalent to 1.5 ppm TOC (the TOC limit for the protocol). If I don't measure within the expected range for the control values, this suggests something is wrong. Of course, if I were to run the 1.5 ppm KHP solution as a control, I may also have to run a blank (depending on whether my control range was based on the value in

DOI: 10.1201/9781003366003-5

the solution itself or the value corrected for the blank). The advantage of running a blank in this situation is that if my KHP control is not correct, I don't know whether the source is the water I used or something else, making my investigation into the root cause more complicated.

For the use of controls, I might further define both *negative* controls and *positive* controls. The negative control is something that has no or very low response. In the TOC example given, the analysis of the water alone (well, alone in a vial) would be a negative control. The use of the 1.5 ppm KHP would represent a positive control, because it would represent a significant analytical response, in line with what could be measured in my protocol (although if my test sample limit were 1.5 ppm TOC, I would hope that my test samples were significantly below that concentration).

Now we get to the complicated part. A sample that is run (in addition to my test samples) may serve *both* as a blank *and* as a control. The TOC situation illustrates this possibility. As covered above, a TOC blank for swabbing may be a sample prepared containing carbon from the vial, the water, the swab, and the atmosphere it is prepared in. As a blank, the analytical TOC value is subtracted from any swab test sample to determine the TOC *sampled from the surface*. However, that same blank sample may also serve *functionally* as a control. How? Well, based on historical data the combination of the vial, water and swab has generally given values in a certain range (for example, 100–200 ppb TOC). If the blank value is outside that range, I should be careful about accepting any analytical data on my test samples. Here is an obvious example. Let's say my swab blank was 350 ppb and my test sample was 125 ppb. Do I subtract 350 ppb from 125 ppb and say that my test sample was below the detection limit? Or, do I look at those results and say something is wrong with that high blank? Perhaps the samples were labeled wrong, and the test sample was 350 ppb and the blank was 125 ppb. If that situation were actually the case, then the net TOC would be 225 ppb TOC, which may or may not be a passing result. So, in this situation, one sample functioned both as a blank and as a control. It is the *same* sample, but it functions in two different ways.

You might also be thinking of the 1.5 ppm KHP, which I indicated was a positive control, and saying that this should be called a *standard*. In the illustration given, that KHP is not strictly speaking a standard. Yes, I am preparing a known concentration. But the measurement of that solution is not functioning as a standard. A standard is a sample of known concentration that I use for *calibration* purposes. For example, if I were preparing a calibration curve, I would have known standards and measured the response of those standards. However, my assumption is that the analytical response of the standards is correct. Another example would be a pass/fail analytical procedure, where I run both a standard (at the protocol limit) and the test sample; if the response of the test sample were less than the response of the standard, then I have confirmed that the test sample has a concentration less than the concentration of the standard. Thus, the difference between a control and a standard is how the analytical result is used. In the first case (the control), the analyst determines whether the result is acceptable or not.

In the second (the standard), the result of the measurement is not questioned; it is accepted as correct.

Let me reiterate that usages of these terms may be different. The important thing is that they are defined and used appropriately. But it is important to think about the *function* of a sample for use as a blank and/or for use as a control.

The above chapter is based on a *Cleaning Memo* originally published in December 2018.

REFERENCES

Vitha, M.F., Carr, P.W. and Mabbott, G.A. "Appropriate Use of Blanks, Standards, and Controls in Chemical Measurements". *J. Chem. Educ.* 2005, 82(6), pp. 901–902, June 1.

5 Meaning of "Dedicated"?

A common question directed to a cleaning validation SME is how to handle cleaning validation for dedicated equipment. As focused on in these chapters, words and wording are very important when we read regulatory documents or ask questions about what we should do. What "dedicated" means is a good example of this.

I generally see the word "dedicated" to be used in at least one of three ways:

In one case, "dedicated" means I am only making a certain *type of product* on my equipment. For example, the equipment is dedicated to making only dermatologicals. Or it is dedicated to biotech. Or it is dedicated to only making products with highly hazardous actives. Or it is dedicated to materials for clinical trials (investigational medicinal products, or IMPs).

A second case is when "dedicated" means I am only making *one active* on this equipment, albeit *at different strengths*.

A third case is when "dedicated" means I am only making *one formulation* (same active and same excipients, but perhaps with different batch sizes) on this equipment.

I'm sure you can see how each of these may be handled differently for cleaning validation programs. I'm not sure how the first use helps a lot. However, it is a valid use of the term "dedicated". Certainly knowing that only a certain type or class of products is made on my equipment can help me restrict what I have to do for cleaning validation. In all cases, I am still going to set acceptance limits, determine appropriate sampling, decide on appropriate analytical methods, and the like. However, key issues for dealing with acceptable practices will typically be different in biotech facilities, in dermatological facilities, in highly hazardous facilities, and in clinical trial material facilities. While we can discuss what is appropriate in each of those dedicated situations, the fact that it is "dedicated" does not come into play as such when I am trying to set up a cleaning validation program. What is more critical is what it is dedicated *to*.

So, let's move on to the second case, where only one active is made on my equipment. Is this a case of "dedication" that is meaningful for cleaning validation? Does this fit into the FDA statements on "dedication"? Well, to be perfectly clear, the 1993 FDA cleaning validation guidance only uses the term "dedicate" or "dedicated" for two situations where the cleaning is difficult [FDA, 1993]. The first situation involves the following specific sentences in Section III:

Bulk pharmaceutical firms may decide to dedicate certain equipment for certain chemical manufacturing process steps that produce tarry or gummy residues that are

DOI: 10.1201/9781003366003-6

difficult to remove from the equipment. Fluid bed dryer bags are another example of equipment that is difficult to clean and is often dedicated to a specific product.

The second situation involves bulk manufacture where there may be by-products from the manufacture of "potent" actives. The specific wording in Section V is:

> In a bulk process, particularly for very potent chemicals such as some steroids, the issue of by-products needs to be considered if equipment is not dedicated.

Other phrases in that 1993 guidance that may be interpreted (or perhaps misinterpreted) as "dedication" deal with *different batches of the same product*. The specific wording in Section III is:

> If firms have one cleaning process for cleaning between different batches of the same product and use a different process for cleaning between product changes, we expect the written procedures to address these different scenario.

The specific wording in Section IV is:

> Determine the number of cleaning processes for each piece of equipment. Ideally, a piece of equipment or system will have one process for cleaning, however this will depend on the products being produced and whether the cleanup occurs between batches of the same product (as in a large campaign) or between batches of different products. When the cleaning process is used only between batches of the same product (or different lots of the same intermediate in a bulk process) the firm need only meet a criteria of, "visibly clean" for the equipment. Such between batch cleaning processes do not require validation.

The important issue for these two statements is what "batches of the same product" means. If I have a drug product with a level of 100 mg active and a second one with 200 mg of the same active, are those two products the *same* product? I suspect it is a stretch to make that claim when the formulations are different. Yes, there may be ways to simplify cleaning validation in a situation where only those two formulations (different active levels) are the *only* products made on that equipment. However, I would not consider that situation as "dedicated" equipment for cleaning validation purposes. Even though this is not strictly dedication by the third definition above, one simplification in this case may be to use a grouping (matrixing) approach, selecting the higher strength as the "most difficult to clean" product (based on the expectation that it is more likely to leave higher residue levels of the active on cleaned equipment surfaces).

This brings us to the third case, where dedication is one formulation only. Clearly in this case, the concerns for cleaning validation are reduced. This may be a situation where, between batches of the same formulation, I *may* only do what is sometimes referred to as "minor cleaning". Minor cleaning may involve vacuuming between batches of the same formulation in solid oral drug product processing, or a water flush between batches of the same formulation in liquid oral drug product manufacture (or also a solvent flush between batches of the same intermediate or active in small molecule API synthesis).

In this situation, minor cleaning is *not* generally a validated cleaning process. There is not necessarily an expectation that the equipment will be visually clean after such "minor cleaning". The goal of minor cleaning is to minimize batch intermingling and/or to improve process efficiency (such as being able to produce more batches in solid oral dose manufacturing before buildup of product on equipment surfaces interferes with product *physical* properties). While that minor cleaning does not require validation itself, that minor cleaning should be considered as *part* of my overall validation strategy at the end of a fixed number of batches in a campaign of the same formulation. That is, does the "difficulty of cleaning" of that final batch change as I increase the number of batches in a campaign, even though the next campaign may be the same formulation? If the next product in the next campaign is the same formulation, I may also be able to only set limits for my cleaning agent and for bioburden (in non-sterile manufacturing), and strictly rely on visually clean for carryover of the active. This also assumes that there is no buildup of degradation products as the campaign proceeds.

As has been suggested, this concept of "between batches of the same product" may also apply to campaigns where campaigns of different products are made on the same equipment (note that this is taking us outside the concept of equipment dedicated to one formulation). For example, I make eight batches of Product A on equipment, with minor cleaning between batches and a full, validated cleaning after the eighth batch. I then make ten batches of Product B on that cleaned equipment, with minor cleaning between batches and a full, validated cleaning process after the tenth batch. While not strictly "dedicated" equipment, it appears to fall within the constraints given by the FDA of "between batches of the same product".

As a side issue, I will address another issue along this line because the question will inevitably come up. The FDA 1993 guidance also states:

> When the cleaning process is used only between batches of the same product (or different lots of the same intermediate in a bulk process) the firm need only meet a criteria of "visibly clean" for the equipment. Such between batch cleaning processes do not require validation.

I must admit I have never fully understood the intent of this wording. As ordinarily done in the industry, there is generally *no expectation* that minor cleaning between batches of the same product be validated. For example, in solid oral dose manufacture, it generally is not an expectation that equipment be visually clean after minor cleaning. Furthermore, for liquid oral dose manufacture using a water rinse between batches, the equipment may appear to be visually clean when viewed in the wet state (it may be viewed in the wet state if drying is not done). However, it is likely in some cases that equipment viewed in the wet state and noted as visually clean would *not* be visually clean if viewed in a dry state. Perhaps the intent of the FDA is that this requirement for visually clean only applies if a full cleaning process (and not a minor cleaning process) is performed between batches. In that case, visually clean may be adequate to demonstrate a lack of cross-contamination of the active between batches (in this case, contamination refers to changing the concentration of the active in the next processed batch). If that is the case, it would still seem that in many cases, validation would be required if cleaning agents were used or if there were bioburden concerns (so residues of the cleaning agent and/or bioburden would be required). So, I am still not entirely clear about the FDA's intent.

Whatever their intent, my intent here is just to help clarify different uses of the terms "dedicated" and "dedication" so that we understand their different uses, making sure all involved in a discussion are on the "same page".

The above chapter is based on a *Cleaning Memo* originally published in October 2017.

REFERENCES

FDA, *Guide to Inspections Validation of Cleaning Processes*, United States Printing Office, Washington, D.C., 1993.

6 Words (Again?)

I have written several chapters about the importance of using terminology more carefully. While it may seem to be more of a problem now (perhaps because of the widespread dissemination of information on the internet), it is not something new in the arena of cleaning validation. This chapter will focus on two examples from the year 1993, the year sometimes considered the "birth" of cleaning validation, although clearly cleaning validation was "conceived" much earlier [Harder, 1984; Mendenhall, 1989].

The first is the publication by two scientists at Lilly (Gary Fourman and Michael Mullen), published in the April 1993 issue of *Pharmaceutical Technology* [Fourman and Mullen, 1993]. It provided the fundamental basis for calculating limits for pharmaceutical cleaning validation by proposing limits on a carryover basis of the more stringent of 0.001 of a dose and 10 ppm, as well as requiring that the equipment be visually clean. The dose criterion is given as "No more than 0.001 dose of any product will appear in the maximum daily dose of another product". The 10 ppm criterion is given as "No more than 10 ppm of any product will appear in another product". You might ask, "What's the problem with that?"

The problem is that some people have read (or interpreted) that proposal as meaning a criteria referring to limits of one *drug product* in another drug product. So, when they take a look at the calculation, they base the dose criteria on the dose of the cleaned drug product in another drug product, or base the 10 ppm criterion on no more than 10 ppm of any drug product in another drug product. It should have been clear from the examples the Lilly scientists used that the residue of concern was the active (API) in the cleaned drug product. In one example given in that publication, the factor "I" related to the dose of the cleaned product is based on "milligrams of the active ingredient". That is, the calculation was not a "0.001 of a minimum dose of one drug product in another drug product", but rather was "0.001 of a minimum dose of the active of one drug product in the maximum dose of another drug product". That is, rather than using the term "product" loosely as meaning either a "drug product" or an "active ingredient", it would have been clearer to use more precise terms (as previously discussed in Chapter 1 of this volume).

This also carries over to the use of the 10 ppm criterion. It is not a 10 ppm drug product in another drug product. Again, in the example given for the 10 ppm criterion, the term "R" is "10 mg of the *active ingredient* in product A/kg product B" [emphasis added]. So, the calculation is not "10 ppm of one product in any other product" but rather 10 ppm of the active of one drug product in any other drug product.

Don't get me wrong here and think I'm complaining about that publication. That paper was critical in the early days of cleaning validation for providing a

DOI: 10.1201/9781003366003-7

basis for how limits could be set. A careful reading of the paper should allow a reader to clearly know when the "product" referred to was the "active ingredient" and when it was actually the "drug product". I am merely using this as an example to encourage us to be more careful in the use of terms.

The second example I will cover is also from 1993. It is the FDA's cleaning validation guidance document [FDA, 1993]. And this example also covers the issue of limits. In that document, the FDA state that "limits that have been mentioned by industry representatives … include analytical detection levels such as 10 PPM, biological activity levels such as 1/1000 of the normal therapeutic dose, and organoleptic levels such as no visible residue". It is likely that those three "industry examples" are from the Fourman and Mullen paper since that publication is listed in the "references" at the end of the FDA guidance. Yes, the FDA presentation is slightly different in that it refers to the "normal therapeutic dose" as opposed to the minimum therapeutic dose, as well as expanding visually clean to *other organoleptic evaluations*.

However, what I want to focus on here is the FDA's reference to 10 ppm as the analytical detection level. As an analytical detection level, this sounds like it means that as long as the analytical sample test result is below 10 ppm, any measured residue is acceptable. We all know this can't be the case; if it were, we would just extract our swab samples with a much larger volume of diluent to make sure the measured value was below 10 ppm. But that is *not* what is stated in the Fourman/Mullen paper. The 10 ppm is 10 ppm as the concentration in the next drug product. I suspect that this is part of the reason that I still see people wanting to set limits just based on a 10 ppm value in an analytical sample. For clarification, doing so may be possible, as long as "10 ppm in an analytical sample" was *both* less than a limit calculated on a dose criterion *and* less than a limit calculated on a "10 ppm in the next product" criterion.

So, is there a practical point to giving these examples? Of course! The first is to never use the word "product" in a generic sense unless you clearly state you are doing so. For example, sometimes I will write "As used in this report, 'product' may mean drug product, active, excipient or cleaning agent". But, if I am specifically referring to the formulated product, I try to refer to it as the "drug product" (or sometimes as formulated drug product, to be somewhat redundant). And I refer to the active as drug substance, API, or drug active. It may seem wordy, but it avoids potential confusion. Second, when you refer to a limit make sure you state what that limit is a limit in (or what it is a limit for). There is a clear difference between 10 ppm in the next product and 10 ppm in the extracted swab sample. This is one of the reasons that I try to use the "L0, L1, L2, L3, L4a, L4b, and L4c" terminology [LeBlanc, 2013] when I discuss limits.

Some of you may tire of me talking about the importance of clear expressions of terminology and definitions. However, providing that clarity in your thinking will only help you design and execute a better cleaning validation program.

The above chapter is based on a *Cleaning Memo* originally published in May 2020.

REFERENCES

FDA, *Guide to Inspections Validation of Cleaning Processes*, United States Printing Office, Washington, D.C., 1993.

Fourman, G.L.; Mullen, M.V. "Determining Cleaning Validation Acceptance Limits for Pharmaceutical Manufacturing Operations". *Pharm. Tech.* 1993, 17, pp. 54–60.

Harder, S.W. "The Validation of Cleaning Procedures". *Pharm. Tech.* 1984, 8, pp. 29–34 (May).

LeBlanc, D.A., "My Revised Shorthand for Expressing Limits", in *Cleaning Validation: Practical Compliance Solutions for Pharmaceutical Manufacturing*, Volume 3. Parenteral Drug Association, Bethesda, MD, 2013, pp. 69–71.

Mendenhall, D.W. "Cleaning Validation". *Drug Dev. Ind. Pharm.* 1989, 15(13), pp. 2105–2114.

Section II

Health-Based Limits

The following nine chapters cover issues related to ADE and PDE values as advocated by Risk-MaPP and by EMA.

DOI: 10.1201/9781003366003-8

7 What's at Stake with HBELs

It might seem like ancient history, but in December 2016, the EMA issued a *draft* "Q&A" [EMA 2016] clarifying some issues in its 2014 guidance on limits in shared facilities [EMA 2014]. I for one thought it was a "breath of fresh air". In June 2017, the EMA had a stakeholders' meeting where a variety of scientists from regulatory agencies and industries were invited to present their views on the Q&A document [EMA, 2017]. The presentations at that meeting included the expected ones objecting to the EMA "backtracking" (my term) by allowing for the traditional approach of 0.001 of a dose for products that were not highly hazardous, as well as ones describing the uncertainties and confusion that has occurred as a result of the 2014 guideline. The EMA had indicated they intend to issue a clarified Q&A by the end of the year. That "Q&A" document was finalized and issued in 2018 [EMA, 2018]. Later chapters in this volume will cover in more detail what was in the 2016 *draft* Q&A (Chapters 9 and 10) as well as the changes made for the final 2018 *final* Q&A document (Chapters 11, 12, and 13). The focus of this chapter is why these Q&A clarifications were needed because of the lack of specificity in the original 2014 guidance. Does it focus on the issue of what is at stake if the EMA truly back-pedals on the use of the dose criterion for non-highly hazardous products? In one sense it may be a moot point for me since I currently advocate selecting the most stringent of a dose-based limit, an HBEL limit, and a 10 ppm (in the next product) limit [LeBlanc, 2017a]. Using that approach, it is generally the case that the HBEL limit will not be the most stringent criterion for non-highly hazardous products, but that the HBEL will generally be the most stringent limit for highly hazardous products. However, it is still worthwhile to understand and evaluate the thinking process in going from the 2014 guidance to the 2016 draft Q&A to the 2018 final Q&A.

Before I get to what's at stake, I would like to suggest why the EMA 2016 Q&A document is *not inconsistent* with the 2014 guideline. Everyone (okay, that's an exaggeration) appears to focus on the requirement in the 2014 guideline that PDEs (Permitted Daily Exposures) be used as a health-based exposure limit (HBEL). As I have previously pointed out [LeBlanc, 2017b; LeBlanc 2017c; LeBlanc 2017d], the 2014 guideline (in Section 4.2) has the following statement

DOI: 10.1201/9781003366003-9

about products where the clinical data (not animal studies) are the primary basis for evaluating patient safety:

> If the most critical effect identified to determine a health-based exposure limit is based on pharmacological and/or toxicological effects observed in humans rather than animals, the use of the PDE formula may be inappropriate and a substance-specific assessment of the clinical data may be used for this purpose.

Now, there might be different ways of interpreting this statement and applying relevant clinical data, but at least one way is the approach specified in the 2016 Q&A draft, that for products that are not highly hazardous, the use of the "traditional" approach of 0.001 of a dose and 10 ppm is adequately protective of patients and constitutes an HBEL for those products that are not highly hazardous.

Note that there is another section in the 2014 guideline where PDE *may* not be appropriate. In Section 5.3 is the following statement relating to biotech manufacture:

> Therapeutic macromolecules and peptides are known to degrade and denature when exposed to pH extremes and/or heat, and may become pharmacologically inactive. The cleaning of biopharmaceutical manufacturing equipment is typically performed under conditions which expose equipment surfaces to pH extremes and/or heat, which would lead to the degradation and inactivation of protein-based products. In view of this, the determination of health based exposure limits using PDE limits of the active and intact product may not be required.

Obviously in this situation, the use of the 0.001 dose criterion of the native active doesn't necessarily fit. Approaches such as using estimated PDE values of the degraded/inactivated fragments (see Chapter 23 of this volume) could be used (although the EMA was silent on what approaches might be used).

A final note on this issue in the 2014 guideline involves investigational medicinal products (IMPs) where there is *inadequate* information to establish a PDE (see section 5.5 of this 2014 guideline). In this section, the EMA provides options of using a "one size fits all" type of approach using something like a TTC, where there are "fixed" acceptable values based on a toxicologist's determination of the level of hazard.

So far, what I have said is just an introduction to point out that what is given in the 2016 Q&A is *not inconsistent* with what was in the 2014 guideline. Now to my point as to what's at stake.

What is at stake is the difference between an HBEL and a cleaning validation limit (CVL). That is, an HBEL should be considered in setting a CVL; the CVL should be *at least as stringent as* the HBEL. However, a CVL should also consider *other* effects, such as the effect on the quality of the next product. In other words, in addition to patient safety when a residue is carried over into a subsequent product, firms should also consider other possible effects of the residue. This idea of considering effects on quality is not something new. The 2015 FDA Q&A on

CGMP [FDA, 2015], a rewrite of an earlier 1995 Human Drug CGMP Note [FDA, 1995] states the following:

> Equipment should be as clean as can be reasonably achieved to a residue limit that is documented to be safe, causes no product quality concerns, and leaves no visible residues.

It is possible to go back even further and find the statement below from a 1992 publication by toxicologists from Abbott [Conine et al., 1992]:

> In practice, the actual allowable residue concentration in a pharmaceutical should be based upon both health and product quality concerns. Thus, the residue limit(s) derived from this procedure may not always be the binding constraint on an allowable residue concentration for a residue in a pharmaceutical. For example, if a residue limit were 1 mg per day and the maximum daily dose of the pharmaceutical were 10 mg per day, the residue could potentially make up a significant fraction of a daily dose without harming the patient. Obviously, a residue present at such concentrations would not be acceptable. In these cases, the allowable residue concentration should be controlled by product specifications, good manufacturing practices, or other quality-based requirements, and not by the health-based residue limit, so long as the health-based residual limit is always met.

In other words, there may be issues other than patient safety that should be considered in setting CVLs, provided that the patient safety issues are adequately addressed.

Now, if we only set limits based on an HBEL, how can we determine what additional factors may make the limits more stringent to deal with any detrimental effects on product quality? Do we have to start doing studies that look at the effects of residues on product stability or on active bioavailability? If we insist only on HBELs, this is certainly something that we should consider, and something where regulators could ask for rationales and/or data to address those other product quality concerns that may *not* be addressed by a toxicology assessment. I had hoped we would not go down this road, but there are still those who insist that limits are based on HBEL (ignoring the dose-based calculation and the 10 ppm) in the next product criterion [ISPE 2017]. My view on the 10 ppm of residue (in the next product) as an alternative limit to use *if it is more stringent* than the 0.001 dose criteria (as given in the 1993 Fourman and Mullen publication [Fourman and Mullen, 1993]) is that the 10 ppm alternative should, in most cases, address other quality effects. For clarification, this is not the rationale given in the Fourman and Mullen paper, but is my view as to its *true* significance. Note that in the 2016 EMA *draft* Q&A this 10 ppm is also a requirement for highly hazardous actives, that is, highly hazardous products have to meet the PDE requirement, but should also meet the *traditional* criteria of 0.001 of a dose and 10 ppm.

So, this discussion of setting limits is not just a matter of what's right (although my opinion is that for the most part the 2016 *draft* Q&A got it right). It is also a matter of what the impact of any change might be.

Let me also clarify that the 2016 EMA *draft* Q&A still leaves the toxicologists and pharmacologists in the driver's seat for setting HBELs. They are the ones to determine if the active is highly hazardous, *and* whether the traditional approach provides adequate patient protection for those that are not highly hazardous.

Finally, while I make a distinction between patient safety and product quality issues, realize that product quality issues *may* affect patient safety and/or health. For example, if a residue shortens the shelf life of a drug product, the patient might not obtain the beneficial outcome expected from the use of the drug product.

The above chapter is based on a *Cleaning Memo* originally published in December 2017.

REFERENCES

Conine, D.A., Naumann, B.D., and Hecker, L.H. "Setting Health-Based Limits for Contamination in Pharmaceuticals and Medical Devices". *Quality Assurance: Good Practice, Regulation, and the Law*, 1992, 1(3), pp. 171–180, June.

EMA/410936/2017 13 July 2017. https://www.ema.europa.eu/en/documents/minutes/summary-discussions-workshop-generation-use-health-based-exposure-limits-hbel_en.pdf (accessed May 7, 2021).

European Medicines Agency. "Guideline on setting health based exposure limits for use in risk identification in the manufacture of different medicinal products in shared facilities". Document EMA/CHMP/CVMP/SWP/169430/2012. London. 20 November 2014.

European Medicines Agency, "Questions and answers on implementation of risk based prevention of cross contamination in production and 'Guideline on setting health based exposure limits for use in risk identification in the manufacture of different medicinal products in shared facilities' (EMA/CHMP/CVMP/SWP/169430/2012)". Document EMA/CHMP/CVMP/463311/2016 (15 December 2016). http://www.cleaningvalidation.com/files/128455599.pdf (accessed May 3, 2021).

European Medicines Agency. "Summary of discussions at the workshop on the generation and use of health-based exposure limits (HBEL) held on 20–21 June 2017 at the European Medicines Agency (EMA)". 13 July 2017

European Medicines Agency, "Questions and answers on implementation of risk based prevention of cross contamination in production and 'Guideline on setting health based exposure limits for use in risk identification in the manufacture of different medicinal products in shared facilities' (EMA/CHMP/CVMP/SWP/169430/2012)". Document EMA/CHMP/CVMP/246844/2018. 19 April 2018.

FDA. "Policy Questions on Cleaning Validation: FDA, Human Drug CGMP Notes". June 1995. http://www.cleaningvalidation.com/cgmp-notes.html (accessed May 3, 2021).

FDA. "Questions and Answer on Current Good Manufacturing Practices – Equipment". Question #7, June 8, 2015. https://www.fda.gov/drugs/guidances-drugs/questions-and-answers-current-good-manufacturing-practices-equipment (accessed May 4, 2021).

Fourman, G.L.; Mullen, M.V. "Determining Cleaning Validation Acceptance Limits for Pharmaceutical Manufacturing Operations". *Pharm. Technol.* 1993, 17, pp. 54–60.

ISPE. *Risk-Based Manufacture of Pharmaceutical Products* 2nd Edition, ISPE, North Bethesda, MD, July 2017.

LeBlanc, D.A., "Shortcomings of ADE/PDE Values for Cleaning Validation", in *Cleaning Validation: Practical Compliance Solutions for Pharmaceutical Manufacturing, Volume 4.* Parenteral Drug Association, Bethesda, MD, 2017a, pp. 23–27.

LeBlanc, D.A., "EMA on Limits for Shared Facilities: Part 1", in *Cleaning Validation: Practical Compliance Solutions for Pharmaceutical Manufacturing, Volume 4.* Parenteral Drug Association, Bethesda, MD, 2017b, pp. 49–53.

LeBlanc, D.A., "EMA on Limits for Shared Facilities: Part 2", in *Cleaning Validation: Practical Compliance Solutions for Pharmaceutical Manufacturing, Volume 4.* Parenteral Drug Association, Bethesda, MD, 2017c, pp. 55–59.

LeBlanc, D.A., "EMA on Limits for Shared Facilities: Part 3", in *Cleaning Validation: Practical Compliance Solutions for Pharmaceutical Manufacturing, Volume 4.* Parenteral Drug Association, Bethesda, MD, 2017d, pp. 61–64.

8 A Look at the Revised Risk-MaPP

In 2017, ISPE released a revised and updated version of its 2010 Risk-MaPP document [ISPE, 2017; ISPE, 2010]. The reasons for the revisions given in the document's preface and in section 1.3 were to address new EU GMP requirements [European Commission, 2015], the EMA guideline on shared facilities [EMA, 2014], and ICH M7 [ICH, 2015], as well as to provide "additional information for cleaning, HVAC, and examples to assist the reader". This chapter focuses exclusively on the changes (and the lack of changes) related to setting limits for cleaning validation purposes.

SOME PARTIALLY "GOOD" NEWS

Unlike the 2010 Risk-MaPP, the 2017 document does not call the 0.001 dose criterion "not science-based". It merely refers to the dose criterion as the "traditional" approach (reflecting the same terminology given by the EMA in its recent *draft* Q&A [EMA, 2016]). However, it still is not the "preferred" strategy (section 6.3.2.3). It might be possible to interpret this as meaning the "traditional" approach is an acceptable strategy, just not preferred. However, that is not explicit in the updated document. If ISPE views the dose criterion as an acceptable, but not preferred strategy, then it would be helpful if it clearly stated that it is only an acceptable strategy for non-highly hazardous actives, much like what is stated in the *draft* EMA Q&A. This is a step in the right direction. That said, the revised Risk-MaPP still refers to the ADE approach as the "science-based" approach, in contrast to the traditional approach. What I don't quite seem to understand is why the "traditional" approach is not considered a science-based approach. After all, in the *draft* EMA Q&A, the traditional approach is considered an HBEL for products that are not highly hazardous. Are we to conclude that the EMA is advocating something that was not science-based? Finally, in discussing the traditional approach, ISPE fails to clarify that the traditional approach utilizes the *more stringent* of the 0.001 dose criterion and 10 ppm in the next product. Note that this failure to *fully* describe the traditional approach is also present in the EMA *draft* Q&A. It baffles me why so fundamental a fact is not accurately portrayed.

In either this revised Risk-MaPP or in the 2010 edition, there is no reason given to explain what about the dose criterion prevents it from being called "science-based". I suspect that the objection to the dose criterion is that it is not as specific to the variations that are present in the different drug products. In this sense, a detailed calculation like given for ADEs and PDEs [ISPE, 2017; ICH, 2016] has the appearance of being more "scientific" as compared to the "one size fits all" approach with the dose-based calculation. (Note that when I refer to the traditional dose calculation as being "one size fits all", I mean that the 0.001 dose and 10 ppm criterion is applicable to all products where the primary safety/health concern is the therapeutic effect.)

It seems inconsistent for the Risk-MaPP authors to be concerned about a "one size fits all" approach for cleaning limits, but apparently welcome it in toxicological evaluations. For example, in its Table 5.2 on adjustment factors both Risk-MaPP and the EMA list a factor of 10 as the intraspecies adjustment factor – a "one size fits all" factor. Is it not possible that the intraspecies factor might vary based on the chemical species, the nature of the critical effect, and the nature of the relevant population or sub-population? Now, as has been pointed out to me, I am not a toxicologist, but I assume that this adjustment factor is appropriate based on the judgment of qualified toxicologists. But, my point is that this is *not conceptually different* from the 0.001 dose criteria being applicable to a certain class of products (namely those that where the primary safety concern is the therapeutic effect).

Another example of this inconsistency is the reference in section 5.3.5.3 of an article published in 2005 [Dolan et al., 2005] for setting limits with limited toxicity data. That article lists a three-tiered approach to selecting ADE values (1 μg for products likely to be carcinogens, 10 μg for products likely to be toxic or potent, and 100 μg for others not in the first two categories). Obviously, the Risk-MaPP authors don't consider this to be unscientific but see it as an acceptable strategy.

A second reason that the dose criterion is not preferred for non-highly hazardous products may be that in most cases (probably in almost all cases for non-highly hazardous products) the use of an ADE results in a higher (that is, *less stringent*) limit. This revolves around what Risk-MaPP calls the "Margin of Safety" and the ability to set higher limits. This issue is discussed in more detail in Chapter 9. However, it should be noted (and should be common sense in setting cleaning validation limits) that patient safety is not the sole criterion for setting limits. Other effects on product quality should also be considered (see Chapter 7 of this volume for more on that issue). ISPE seems to ignore other quality concerns in its statement in section 6.3.2.3 that the "only criteria necessary for a robust cleaning process are the health-based, ADE derived limit, a validated analytical method with a sensitivity below the acceptance limit, that is visually clean".

WHAT'S NEW OR DIFFERENT

Sections 3.2.1.4 and 6.3.2.2 discuss compounds that are denatured, degraded, and/or deactivated during the cleaning process. In this situation, an ADE should be developed for the resultant residues.

In section 5.2 is a discussion of using OEL and OEB values to estimate ADE values, mainly for the purpose of "prioritization in risk assessments".

In section 5.3.1 is a description of a "qualified toxicologist". Note that this was a question side-stepped to a large extent by the EMA in Question #9 in its *draft*

Q&A document, but was addressed in the EMA final Q&A document in 2018 [EMA, 2016; EMA, 2018].

In section 5.3.5, the use of the term "uncertainty factors" in ADE calculations has been modified in favor of calling them "adjustment factors". Table 5.2 gives a comparison of adjustment factors in Risk-MaPP and in the 2014 EMA guidance.

In section 5.3.5 is the statement "Ideally, the ADE is based on the route that it will be applied to in the evaluation". This issue is further addressed in section 5.3.5.1, and seems to be saying that if only one route (such as oral) is probable, then ADE values could be adjusted but should be done in a "later step in the risk assessment process". This sounds like it might mean that ADEs can be developed as route-specific, but that a formal ADE by any/all routes of administration should be first determined, and then at a *later step* an adjustment can be made to consider only one route. That is, if the cleaning validation "protocol" (I assume that is what the word "evaluation" applies to) is for an oral route, then an oral ADE can be developed. This emphasis on the primacy of "any/all" routes appears to be based on the description of an ADE in section 2.5.1 as well as the definition of an ADE in section 16.2, which refers to a value that is adequately protective by "any route" and by "all routes".

Section 5.3.5.5 discusses beta lactams. While pointing out that most regulatory agencies have strict requirements for making these in dedicated facilities, ISPE seems to suggest that in jurisdictions where beta lactams can be "co-manufactured" with other products, ADE values should be developed for risk assessments and for setting limits.

Section 5.5 outlines a format for documenting the determination of an ADE.

Section 6.3.2.1 refers to the fact that swab limits based on an ADE may "actually permit the equipment to look dirty ... so therefore the acceptance criteria would be set to visually clean". It is unclear (at least to me) whether the intent here is that a high calculated acceptance limit means that a visual clean criterion could be the *only* criterion for a cleaning validation protocol. Section 6.3.2.3 refers to the use of ADE limits allowing for visual detection for monitoring as part of routine operations (after completing protocols). Note that visually clean is also discussed in section 6.3.2.10.

Section 6.3.2.1 seems to call for establishing microbial limits based on a "very similar procedure" as done for chemical species. No further elaboration or detail is given on how this is actually done.

Sections 6.3.2.7, 6.3.2.8, and 6.3.2.9 deal in more detail (as compared to the 2010 Risk-MaPP version) with introducing new products and with small-scale cleanability studies.

Sections 6.5.2 and 6.5.3 deal with addressing non-product contact surfaces. An example is given of performing a risk assessment by sampling walls and determining the worst-case potential transfer to the next product. Unfortunately, the example given addresses contamination *only* from room walls and does not address the *cumulative* potential transfer of residues from walls *and* from equipment product contact surfaces.

The comments here do not cover all the changes relating to limits, but cover the significant ones for my purposes except for a discussion of the "Margin of Safety" issue, which is covered in Chapter 9.

While I believe the 2017 Risk-MaPP (like the 2010 original document) is *deeply* flawed, for those wanting to implement it for cleaning validation limits, I highly recommend that it be carefully reviewed (as well as reviewing critiques of it) to address its possible implementation.

The above chapter is based on a *Cleaning Memo* originally published in January 2018.

REFERENCES

Dolan, D.G., Naumann, B.D., Sargent, E.V., Maier, A., Dourson, M. "Application of the Threshold of Toxicological Concern Concept to Pharmaceutical Manufacturing Operations". *Regul Toxicol Pharmacol* 2005, 43, pp. 1–9.

European Commission: Directorate-General For Health And Food Safety. EudraLex Volume 4, "EU Guidelines for Good Manufacturing Practice for Medicinal Products for Human and Veterinary Use, Annex 15: Qualification and Validation". Brussels. March 30, 2015.

European Medicines Agency. "Guideline On Setting Health Based Exposure Limits for Use in Risk Identification in the Manufacture of Different Medicinal Products in Shared Facilities" Document EMA/CHMP/CVMP/SWP/169430/2012. London. November 20, 2014.

European Medicines Agency, "Questions and Answers on Implementation of Risk Based Prevention of Cross Contamination in Production and 'Guideline on Setting Health Based Exposure Limits for Use in Risk Identification in the Manufacture of different Medicinal Products in Shared Facilities' (EMA/CHMP/CVMP/SWP/169430/2012)". Document EMA/CHMP/CVMP/463311/2016 (15 December 2016). http://www.cleaningvalidation.com/files/128455599.pdf (accessed May 3, 2021).

European Medicines Agency, "Questions and Answers on Implementation of Risk Based Prevention of Cross Contamination in Production and 'Guideline On Setting Health Based Exposure Limits for Use in Risk Identification in the Manufacture of Different Medicinal Products in Shared Facilities' (EMA/CHMP/CVMP/SWP/169430/2012)". Document EMA/CHMP/CVMP/246844/2018, 19 April 2018.

ICH. "Impurities: Guideline for Residual Solvents". ICH Q3c(R6). October 2016.

ICH M7 "Assessment and Control of DNA Reactive (mutagenic) Impurities in Pharmaceuticals to Limit Potential Carcinogenic Risk". August 2015.

ISPE, *Risk-Based Manufacture of Pharmaceutical Products* (Risk-MaPP), 1st Edition, ISPE, North Bethesda, MD, September 2010.

ISPE, *Risk-Based Manufacture of Pharmaceutical Products* 2nd Edition, ISPE, North Bethesda, MD, July 2017.

9 EMA's Q&A Clarification Part 1

The EMA issued its *draft* Q&A on Health-Based Exposure Limits (HBELs) in December 2106 [EMA, 2016]. There has been surprisingly little attention to this Q&A in the various pharmaceutical journals and internet news sites. My original intention was to wait until a final document is issued before discussing it in detail. However, because the EMA approach in the 2014 guidance [EMA, 2014] was a radical departure from prior industry practices, I addressed it with these comments before it was finalized in 2018 [EMA, 2018]. The comments below should be read both to understand the significant issues that were not addressed in the 2014 guidance, but also to reflect on how this Q&A document changed as it was finalized.

For those of you that are not familiar with this document, it is a series of fourteen (14) questions and answers relating to the implementation of HBELs. Before I start on the 14 questions, a discussion of HBEL vs. PDE values requires some clarification. The more general term is HBEL. PDE is only one method to derive an HBEL. Unfortunately, the 2014 EMA is sometimes read as requiring PDE values for all actives. It should be clear from reading that document that a PDE is only one avenue for establishing an HBEL. That 2014 document also discusses using the TTC concept for genotoxic materials, as well as stating for certain products, like biotechnology actives and products where the most relevant safety data is on humans, the PDE formula *may not* be appropriate.

That said, the wording of the 2014 EMA document may contribute to that misreading. The beginning of Section 4.1 states, "The procedure proposed in this document for determination of health based exposure limits for a residual active substance is based on the method for establishing the so-called Permitted Daily Exposure (PDE)…" Obviously, the PDE is not the *only* procedure to establish an HBEL given in that 2014 document. In addition, Section 6 of the 2014 document is titled "Reporting of the PDE determination strategy"; it would seem appropriate if there were other means of determining an HBEL, that section should have been titled "Reporting of the HBEL determination strategy". Fortunately, the new *draft* Q&A clarifies this and emphasizes the general idea of an HBEL as compared to one embodiment of it (the PDE).

However, this should be a reminder to us all that words and wording can make a difference, and that a *careful* reading of original regulatory documents should always be considered. So take that into consideration as I present a summary and critique of this 2016 draft Q&A. These comments are given for each of the questions as listed in that new draft document. As you read my comments, you should be referring to the EMA draft document at the same time. In addition, care must be used in trying to understand what the EMA means by a "product". In many cases, I believe they are generally referring to a drug active (or drug substance), but in other cases, they are referring to a drug product. Note that the EMA sometimes refers to a "compound", which in this context probably means a drug active.

QUESTION #1

The first question states that "HBELs should be established for all products". That is fairly clear; "all products" is *all* products. However, the answer goes on to state that for "highly hazardous products" the HBEL is "expected to be completed *in full* per the [2104] EMA guide ... or equivalent" [emphasis added]. Question #2 covers highly hazardous products and Question #4 covers those that are *not* highly hazardous.

QUESTION #2

In this question, the EMA defines a highly hazardous product as one that "can cause serious adverse effects at low doses". It lists certain categories (which is not an exhaustive list) of compounds that should be considered highly hazardous. The list includes genotoxic/mutagenic compounds that could be carcinogenic, compounds with reproductive or developmental effects, compounds with specific serious target organ toxicity, highly potent compounds (those with a daily therapeutic dose of <1 mg/day in humans), and compounds that are highly sensitizing.

I also find it difficult to clearly define precisely what I mean by highly hazardous actives. While I am not fully comfortable with the EMA's descriptions, I don't think I can come up with a better list. One area to consider carefully when an assessment is being made using these criteria is the emphasis on effects at "low doses" (<10 mg/day) for mutagenic and reproductive effects. I would think that compounds that were mutagenic should have a full PDE or TTC assessment even if the daily dose was greater than 10 mg/day. I have not yet covered Question #4, but it doesn't make sense to me (as a non-toxicologist) to make that distinction based on the daily dose. There should be a clear distinction between the daily therapeutic dose and the dose at which highly hazardous effects are observed.

The second area of concern is listing "highly potent" compounds, those with a daily dose of <1 mg/day, as highly hazardous. My assumption here is that the EMA is referring to "potent" compounds that are not mutagenic, etc. With other things being equal I would be more concerned about the safety effect of a potent compound as compared to a non-potent compound. But if the major safety concern for the highly potent compound is the therapeutic effect, then the safety concern from 0.001 of a therapeutic dose of a highly potent compound should not be of more concern as compared to 0.001 of a therapeutic dose of a non-potent compound. In my comments to the EMA, I recommended that this be changed.

QUESTION #3

This question addresses the applicability of using OEL or OEB values to "support" an assessment of whether a product is highly hazardous or not. The EMA's answer is "Yes" for determining a *preliminary* PDE. The typical formula of multiplying the OEL/OEB value by 10 m^3 of air is given [Denk, 2017]. The EMA adds that adjustment factors based on the target population and on administration routes may be needed. It also states that PDE values determined in this way that are less than or equal to 10 μg/day should be considered highly hazardous. I am unclear why this last statement is given as to a "highly hazardous" category. My speculation is that the OEL only gives a *preliminary* PDE. If this preliminary PDE is at or below 10 μg/day, then a *full* evaluation should be done to establish the PDE using the approach given the 2014 guidance [EMA, 2014]. If the OEL/OEB approach gives a value above 10 μg/day, then that compound can be considered non-hazardous and the approach given in Question #4 could be used. However, this is *my* speculation of the intent here, and I may be wrong.

QUESTION #4

This is the one I think (or hope) should clearly survive in the final Q&A. This is dealing with products that don't belong to the highly hazardous category and have a "favorable therapeutic index". I think a description of what this means is given later in the answer as compounds where the "pharmacological activity would therefore be the most sensitive/critical effect", with the therapeutic dose being the point of departure for determining the HBEL. Here is the key sentence in full:

> Under these circumstances, HBEL based on the 1/1000th minimum therapeutic dose approach would be considered as sufficiently conservative and could be utilized for risk assessment and cleaning purposes.

This appears to mean that for compounds that are *not* highly hazardous (as defined in Question #2), 0.001 of a dose can be used as the HBEL, and it is not necessary to do a full PDE assessment as given in the 2014 EMA guide.

That said, I believe it would still be prudent to perform a preliminary PDE based on the OEL/OEB if that OEL/OEB data is available.

QUESTION #5

This answer states that an LD$_{50}$ value is *not* an "adequate point of departure to determine an HBEL". My only concern here is that perhaps the context of that statement is a limit for actives. Clearly, for actives, there *has to be* relevant data other than an LD$_{50}$ study. However, for cleaning agents and for intermediates in API synthesis, LD$_{50}$ data may be the only relevant safety data available. I hope that this answer is interpreted only in the context of drug actives. This should be clarified in the final Q&A document.

There are nine more questions and answers to cover. We continue in Chapter 10 with those. Please see Chapters 11, 12, and 13 for how these questions *and* how these answers changed in the final 2018 version of this EMA Q&A document.

The above chapter is based on a *Cleaning Memo* originally published in July 2017.

REFERENCES

Denk, R.; Flueckinger, A.; Kisaka, H.; Maeck, R.; Restetzki, L.; Schreiner, A.; Schulze, R. "Isolator Surfaces and Contamination Risks to Personnel". *PDA Letter*, 6 November 2017.

European Medicines Agency. "Guideline On Setting Health Based Exposure Limits for Use in Risk Identification in the Manufacture of Different Medicinal Products in Shared Facilities". Document EMA/CHMP/ CVMP/SWP/169430/2012. London. 20 November 2014.

European Medicines Agency, "Questions and Answers on Implementation of Risk Based Prevention of Cross Contamination in Production and 'Guideline on Setting Health Based Exposure Limits for Use in Risk Identification in the Manufacture of Different Medicinal Products in Shared Facilities' (EMA/CHMP/CVMP/SWP/169430/2012)". Document EMA/CHMP/CVMP/463311/2016. 15 December 2016. http://www.cleaningvalidation.com/files/128455599.pdf (accessed May 3, 2021).

European Medicines Agency, "Questions and Answers on Implementation of Risk Based Prevention of Cross Contamination in Production and 'Guideline on Setting Health Based Exposure Limits for Use in Risk Identification in the Manufacture of Different Medicinal Products in Shared Facilities' (EMA/CHMP/CVMP/SWP/169430/2012)". Document EMA/CHMP/CVMP/246844/2018. 19 April 2018.

10 EMA's Q&A Clarification
Part 2

This is a continuation of Chapter 9 dealing with the EMA's *draft* Q&A on HBELs for cleaning validation [EMA, 2016]. Please read that chapter before jumping into this one. This chapter covers Questions #6 through #14. As mentioned in the previous chapter, care must be used in trying to understand what the EMA means by a "product". In many cases, I believe they are referring to a drug active (or drug substance), but in other cases, they are referring to a drug product. Note also that the EMA sometimes refers to a "compound", which in this context probably means a drug active.

QUESTION #6

The first five questions covered in the last chapter were about setting HBELs. This question addresses how to set limits for cleaning validation purposes. It states that limits should not be set on a calculated HBEL *alone*, but other factors should be considered. Those other factors include "uncertainty in the cleaning process and analytical variability". Although not stated by the EMA's answer, I would add factors such as effects on product quality and product purity as possibly affect cleaning validation limits, even though those factors are not part of a *health-based* exposure limit. For *non-highly hazardous products*, the EMA states that this can be achieved by "traditional cleaning limits used by the industry such as 1/1000th of minimum therapeutic dose or 10 ppm of one product in another product". This is additional support for *not* requiring a full PDE assessment for these non-highly hazardous products. Although not clearly stated by the EMA, the traditional industry approach for these non-highly hazardous actives has been the *more stringent* of 0.001 of a dose and 10 ppm in the next drug product, *not* an "either/or, choose one".

For highly hazardous products, the EMA also states that limits beyond the HBEL may be appropriate, and then further states that limits for highly hazardous products "should not be higher than the traditional cleaning limits approach". This is apparently a reference to the "traditional approach" mentioned earlier in this Question #6, namely the idea of 0.001 of a dose and 10 ppm in the next product. If this is the case, what it means is that for highly hazardous products, the limit for an active should probably be the *most stringent* of the following three criteria:

1. Calculated HBEL by PDE or TTC criterion for active
2. 1/1000th daily dose of the active in the next drug product
3. 10 ppm of active in the next drug product

QUESTION #7

This answer states that Ectoparasiticides may be manufactured in shared equipment with other human or veterinary products only if supported by HBEL data. This makes sense, and no comment by me is necessary.

QUESTION #8

This question deals with veterinary products for different species manufactured in the same facility. For highly hazardous products with known sensitivity with certain species, the HBEL should take into account "specific animal toxicity knowledge". For products not highly hazardous, the traditional approach (0.001 dose and 10 ppm) mentioned in Question #6 may be used.

QUESTION #9

This question deals with how one determines that the toxicologist performing the HBEL assessment is a *competent* toxicologist. The answer given is to review the person's "experience and qualifications".

QUESTION #10

This question deals with HBELs for early phase IMPs (clinical trial products). The answer given is basically to evaluate "all available data" and to update the assessment as *new* information is available. Furthermore, where knowledge is less than complete it may be appropriate to use the "read across" [ECHA, 2017] approach by evaluating data from similar molecules as well as any other "appropriate" adjustment factors based on worst-case assessments.

QUESTION #11

This question deals with pediatric products that are made in shared equipment/ facilities with products for adults or for animals. The EMA states that the HBEL in this situation should be based not on the adult human weight of 50 kg, but on weights of 10 kg for children, 3.5 kg for newborns, and 0.5 kg for prematurely born newborns. I assume this is mainly for setting limits for the human adult and animal products where residues could potentially transfer to products for the pediatric population. However, those lower weights should also be considered for pediatric products following other pediatric products. [Note that while this "Question" does not appear in the final Q&A document, the weights given for human youth should be considered.]

QUESTION #12

This deals with the question of the relationship of HBELs to the requirements of GMP Chapter 5 section 20 [European Commission, 2015]. This section of the

EU GMPs deals with a Quality Risk Management assessment that "should be the basis for determining the necessity for and extent to which premises and equipment should be dedicated to a particular product or product family". The EMA's answer is that HBELs should be used to *justify* cleaning limits. In this context, it should be remembered that HBEL values are not determined just by PDE or TTC calculations. According to the EMA answers in *this* document, the 0.001 dose criterion also comes into play under the HBEL umbrella; however, the *explicit* statement that 0.001 dose is an HBEL for non-highly hazardous actives does not appear in the final Q&A document.

QUESTION #13

This question deals with the appropriateness of segregating highly hazardous products in a dedicated area (apart from non-highly hazardous products). While such an approach may prevent cross-contamination of the highly hazardous products into the non-highly hazardous products, such segregation *alone* does not address the issue of cross-contamination *between* highly hazardous products. It is still necessary to perform a toxicological evaluation of *each* highly hazardous product to ensure that it does not cross-contaminate another highly hazardous product. In other words, it is *not* adequate to claim that the residue of one mutagenic product in another mutagenic product is not an issue; it *clearly is* an issue.

QUESTION #14

This answer deals with the application of the TTC guide of 1.5 µg/day to mutagenic products as an "acceptable default" approach. The answer given is "Yes" *except* for highly sensitizing compounds (which would include beta lactams). This should not be misread to think the EMA is allowing the TTC approach for *all* highly hazardous products. It is merely saying that the TTC approach is acceptable for mutagenic products, but *not* if those mutagenic products are also highly sensitizing products.

This finishes my observations and comments on the specific questions in this draft document. Two additional comments may be helpful. One is that even though a visually clean criterion is not mentioned in either the 2014 document or in the draft Q&A, it is probably still an expectation that the equipment be visually clean in a product-to-product *validated* cleaning process. The second is that the EMA appears to be making a *distinction* between a health-based exposure limit (HBEL) and a cleaning validation limit. An HBEL only deals with patient safety issues. It *must* be considered in arriving at a cleaning validation limit. However, a cleaning validation limit should consider other relevant factors, such as the 10 ppm criterion (dealing with product purity issues) and the visually clean criterion (dealing with GMP expectations).

While some of the principles in this draft Q&A document may be considered in understanding the EMA's thinking, not all things survived into the final document.

The final document should be consulted and used to support a company's practices.

Please see chapters 11, 12, and 13 for how these questions and how these answers changed in the *final* 2018 version [EMA, 2018] of this EMA Q&A document.

The above chapter is based on a *Cleaning Memo* originally published in August 2017.

REFERENCES

ECHA (European Chemicals Agency), "Read-Across Assessment Framework (RAAF)" Reference ECHA-17-R-01-EN, (2017). https://echa.europa.eu/documents/10162/13628/raaf_en.pdf/614e5d61-891d-4154-8a47-87efebd1851a (accessed May 4, 2021).

European Commission. "EU Guidelines for Good Manufacturing Practice for Medicinal Products for Human and Veterinary Use", Part 1. Chapter 5: Production. Ref. Ares(2015)283689. 23 January 2015.

European Medicines Agency, "Questions and Answers on Implementation of Risk Based Prevention of Cross Contamination in Production and 'Guideline on Setting Health Based Exposure Limits for Use in Risk Identification in the Manufacture of Different Medicinal Products in Shared Facilities' (EMA/CHMP/CVMP/SWP/169430/2012)", Document EMA/CHMP/CVMP/463311/2016. 15 December 2016. http://www.cleaningvalidation.com/files/128455599.pdf (accessed May 3, 2021).

European Medicines Agency, "Questions and Answers on Implementation of Risk Based Prevention of Cross Contamination in Production and 'Guideline on Setting Health Based Exposure Limits for Use in Risk Identification in the Manufacture of Different Medicinal Products in Shared Facilities' (EMA/CHMP/CVMP/SWP/169430/2012)", Document EMA/CHMP/CVMP/246844/2018. 19 April 2018.

11 The EMA Q&A "Clarification" on Limits

The EMA issued its clarifying *final* Q&A on limits in shared facilities on April 19, 2018 [EMA, 2018]. This was a "final" document with changes from the *draft* version issued in January 2016 [EMA, 2016]. I say it was a clarifying document. However, in my view, it just muddied the waters further as compared to the clear statements made in the draft document. That said, it is probably not as bad as it looks at first glance, providing you read it *carefully*. In this chapter and in the next two chapters, I will explain what is being said and what is not being said in this Q&A document. In this chapter, the focus is on the issue of Health-Based Exposure Limits (HBELs) and the traditional way limits were set. As you read what follows, note that some statements will refer to the *final* document and some statements will refer to the earlier *draft* document.

First, the EMA is clear in its answer to Question #1. HBELs are required for *all* "medicinal products". This is a *slight* change from the draft document, which required HBELs for all "products". I might be reading too much into this change, but it might mean that HBELs, as defined in this document, are not required for cleaning agents and for intermediates (in small molecule API synthesis). Now it should be clear that toxicological evaluations should be part of setting limits for cleaning agents and intermediates. However, in many instances, the most relevant toxicity information is short-term toxicity information such as animal LD_{50} values. And, if we jump to Question #10, LD_{50} "is not an adequate point of departure to determine an HBEL for drug products". This has a *slight* twist from the draft document where the final clarifying phrase "for drug products" is *absent*. Does this mean that LD_{50} values can be a starting point for an HBEL for detergents and intermediates (which clearly are *not* drug products)? We might have to wait for a further clarifying Q&A from EMA on this issue.

Second, the answer to Question #2 merely states the obvious that there is a hard continuum (with no cut-off points) in evaluating the level of risk from any hazard. While this should not come as a surprise, it is a change from Question #2 in the *draft* document which made a distinction between defined *highly hazardous products* and those that are *not* highly hazardous. Yes, that distinction does involve a cut-off point, but it is a reasonable cut-off point like those used when a qualified toxicologist establishes OEL bands for worker safety purposes (those bands do have hard cut-off points, although everyone realizes that the distinction between a value of 0.99 and 1.01 may not be significant). Having a hard cut-off helps us make useful distinctions in the real world, as opposed to theoretical (but clearly true) assertions. It would seem (at least to me) that the distinction between an

DOI: 10.1201/9781003366003-13

active where the therapeutic effect is the point of departure and an active where there is a significant *other* effect of the active (giving a different point of departure) is a valid one. We should also remember that for Risk-MaPP and the EMA, the original impetus for their efforts was to find some way to set limits for "certain" actives (which were the highly hazardous ones) where the traditional dose-based criteria were *not* applicable [ISPE, 2010; EMA, 2011].

Third, Question #6 in this final Q&A is "How can limits for cleaning purposes be established?". The first sentence of that answer is the same one given in the draft document: "Although the EMA guideline (EMA/CHMP/CVMP/SWP/169430/2012) may be used to justify cleaning limits (as per Introduction paragraph 3), it is not intended to be used to set cleaning limits at the level of the calculated HBEL". I think what this suggests is that there is a distinction between an HBEL and a cleaning validation limit. My understanding of this is that the HBEL is a requirement to ensure patient safety and the cleaning validation limit should meet this requirement *at a minimum*. However, the cleaning validation limit may be *more stringent* based on other concerns, such as product quality and purity. If my understanding is correct, this is a far cry from the statement in the 2017 Risk-MaPP that "The only criteria necessary for a robust cleaning process are the health-based, ADE derived limit, a validated analytical method with a sensitivity below the acceptance limit, that is visually clean" [ISPE, 2017].

The answer to Question #6 then goes on to talk about "historically used cleaning limits" for *existing* products. It states quite clearly that those limits "should be retained" and that those cleaning limits can be used as *alert* limits to "provide sufficient assurance that excursions above the HBEL will be prevented". That is, it assumed that the historically used limit will be at least as stringent as the HBEL. Now, here is a clarification from me. The historically used limits (which are likely the same as what was called the traditional limits of 0.001 dose and 10 ppm in the *draft* Q&A) have only been applicable for products where the therapeutic effect of the actives is the point of departure for establishing limits; in most cases, for highly hazardous actives, the HBEL will be more stringent.

The statement in the answer to Question #6 about the "historically" used limits and using them to assure that the HBELs will not be exceeded is followed by the statement "A similar process should be adopted when establishing cleaning alert levels for products introduced into a facility for the first-time". This seems to imply that the historically used limits may be used for *new* products in a facility in the *same way* as for existing products. It certainly is not clear to me what the EMA is trying to say; however, my opinion is that the statements in the *draft* Q&A are a better approach for a clarifying Q&A document. Essentially what the draft document implied was that for *all* actives, you should determine the limits based on an HBEL, 0.001 dose, and 10 ppm, and use the most stringent of those three calculated limits. Note that the draft document did not explicitly say to use that for *all* actives. But in reading what is recommended for highly hazardous actives and for non-highly hazardous actives, that conclusion is the ultimate result. My understanding is that for most *highly hazardous actives* either the HBEL or the 10 ppm criterion will be the most stringent, while for most *non-highly hazardous actives*, either the 0.001 dose or the 10 ppm criterion will be the most stringent. That approach is one I have recommended for several years because of my concern that HBELs only consider patient safety in setting limits and ignore other concerns related to product quality.

I think what this means is that if you want to implement the EMA guideline, you should carefully read and understand this new Q&A document, and consider all comments (mine as well as comments from others) about what it *really* states and *really* means. My disappointment in the EMA is that they had the opportunity to clarify their 2014 guideline, but only ended up obfuscating issues relating to limits in what has become a highly political atmosphere in the scientific world.

Chapter 12 continues with discussions of other issues in the final EMA Q&A document.

The above chapter is based on a *Cleaning Memo* originally published in June 2018.

REFERENCES

European Medicines Agency, "Concept Paper on the Development of Toxicological Guidance for Use in Risk Identification in the Manufacture of Different Medicinal Products in Shared Facilities". EMA/CHMP/SWP/598303/2011. October 2011.

European Medicines Agency, "Questions and Answers on Implementation of Risk Based Prevention of Cross Contamination in Production and 'Guideline on Setting Health Based Exposure Limits for Use in Risk Identification in the Manufacture of Different Medicinal Products in Shared Facilities' (EMA/CHMP/CVMP/SWP/169430/2012)". Document EMA/CHMP/CVMP/463311/2016. 15 December 2016. http://www.cleaningvalidation.com/files/128455599.pdf (accessed May 3, 2021).

European Medicines Agency, "Questions and answers on implementation of risk based prevention of cross contamination in production and 'Guideline on setting health based exposure limits for use in risk identification in the manufacture of different medicinal products in shared facilities' (EMA/CHMP/CVMP/SWP/169430/2012)". Document EMA/CHMP/CVMP/246844/2018. 19 April 2018.

ISPE. *Risk-Based Manufacture of Pharmaceutical Products* (Risk-MaPP), 1st Edition, September 2010.

ISPE. *Risk-Based Manufacture of Pharmaceutical Products* 2nd Edition, ISPE, North Bethesda, MD, July 2017.

12 The EMA Q&A on Routine Analytical Testing

As discussed in Chapter 11, the EMA issued it clarifying Q&A on limits in shared facilities on April 19, 2018 [EMA, 2018]. Chapter 11 focused on HBELs and "historically used" limits. In this chapter, the focus is Questions #7 and #8 (Q7 and Q8) dealing with *routine analytical testing*.

The question in Q7 is as follows: "Is analytical testing required at product changeover, on equipment in shared facilities, following completion of cleaning validation?" The answer starts off with the following: "Analytical testing is expected at each changeover unless justified otherwise via a robust, documented Quality Risk Management (QRM) process". That answer continues with three things that should be considered *at a minimum* as part of the QRM process: (a) the repeatability of the cleaning process (manual vs. automated), (b) the residue hazard, and (c) "whether visual inspection can be relied upon to determine the cleanliness of the equipment at the residue limit justified by the HBEL".

It makes sense to do a risk assessment to determine what should be done for analytical testing on routine cleaning after completion of the validation protocols. The key item in the list of three minimum considerations to be covered here is the third one relating to whether visual inspection is adequate to determine meeting the limit calculated using the HBEL. Is that a reasonable expectation? Well, it is a reasonable expectation that the equipment be visually clean at the end of the cleaning process. However, most companies do not assume that being visually clean means you are meeting the calculated limit (whether it is the "historically used" limit based on dosing or a limit based on an HBEL).

The question arises as to whether this is a guideline prescription or recommendation or just an option. If it is something that *must* be done to minimize or avoid analytical testing for routine manufacture, it is adding a whole new layer of testing to support the assertion that the HBEL is being met. After all, the purpose of cleaning validation is to demonstrate that if the cleaning process is carried out correctly, equipment will be cleaned to the applicable residue limits. The proposed "visual limit" type of testing usually involves laboratory spiking studies to determine levels that would indicate passing results in comparison to the calculated limit (typically in $\mu g/cm^2$). It may be possible, however, to avoid laboratory tests if the HBEL-calculated residue limit is *significantly* above values of $1–4$ $\mu g/cm^2$ typically cited for this type of evaluation. That is, if my calculated HBEL limit were to be 15 $\mu g/cm^2$, could I avoid doing spiking studies? While that seems like a reasonable approach, will we then proceed to start arguing about where that cut-off level should be, because we want a "science-based" cut-off level?

Let me also point out that the answer to Question #7 appears to say that visual cleanness must be shown to correlate in a certain way with a limit calculated by using the HBEL, and not necessarily with a cleaning validation limit that may be more stringent than the HBEL limit (consistent with the answer provided to Question #6).

Before we leave this topic, Question #8 and its answer adds another element to this determination of what visually clean means. The question in Question #8 is "What are the requirements for conducting visual inspection as per Q&A 7?" The answer is that "manufacturers should establish the threshold at which the product is readily visible as a residue". This should be interpreted carefully, particularly what the applicability of "threshold" might be. Many that are performing visual limit spiking studies spike with decreasing level of residue to determine the amount that allows for the entire spiked surface to be visually clean, and then use that level (or one level above that level) to establish a "visual detectability limit". I don't think that is a good practice. Why? In any spiking study, it is next to impossible to spike a surface so that the residue is evenly dispersed across the surface. Part of the reason for that is how the residue (in solution) is applied and part is that as the residue dries, not all areas dry at the same rate so there may be some degree of "migration" across the surface. What happens is that *at low levels*, the spiked surface may be visually clean in some or most areas, but will have spots where the residue is clearly visible.

If I spike at nominal levels of 0.1 and 0.2 $\mu g/cm^2$. When I view the 0.1 spike, it is completely visually clean under defined viewing conditions, but when I view the 0.2 spike, I see a few specks of residue scattered on the surface. What does that mean? If my HBEL residue limit is 0.4 $\mu g/cm^2$, does this mean that with a visually clean surface that I am meeting my HBEL limit? My answer is "not necessarily". When I spike the surface at the 0.2 level and only see scattered spots, the concentration of residue *at the locations of those scattered spots* is not 0.2, but is likely somewhat higher. So I would not be using sound science to say that visually clean would mean that residue must be below 0.4. It might be, but might not be.

The approach I usually recommend (see Chapter 26 of this volume) is to spike the surface at a level equal to the calculated limit. In the example given, I would spike the surface at a level of 0.4 $\mu g/cm^2$ and observe it after drying. If it is visibly soiled (that is, not visibly clean) across the *entire* spiked surface, then I could logically conclude that if a surface is visibly clean, it would be at a residue level below that 0.4 value. Note that in this type of evaluation, there would be some areas of the spiked surface that would be below that 0.4 value and some areas above that spiked value. But, since even those lower values are visibly soiled, I can conclude that the amount of the residue would be below that 0.4 value if a surface observed under the *same viewing conditions* in a protocol were visibly clean. I also like this type of evaluation since it is more straightforward. Note that if my residue limit were 0.4, I might do my spiking at 0.3 or even 0.2. If these lower levels also were visibly spoiled across the entire surface, that would be an indication of the robustness of such a visual observation.

Now, I don't think the use of "visibly clean" for routine monitoring *in this manner* is helpful. Don't get me wrong here; it is still appropriate for routine monitoring to perform a visual examination to the extent it is practical. But it should not be required to correlate a visually clean state with meeting the HBEL calculated limit as specified by the EMA answer. For most situations, doing any extensive "dismantling" of equipment (as suggested in the EMA Q&A) to gain access for viewing probably poses an additional and unnecessary risk of adulteration or contamination of the next manufactured product. The situation that should be of most interest for the EMA recommended approach is for highly hazardous actives, and it is precisely those actives which are more likely to have very low residue limits. And with those very low residue limits, it is likely the case that equipment could be visually clean and still have *unacceptable* levels of residues. That is one of the reasons why I would generally recommend for highly hazardous actives that a more appropriate *routine monitoring* approach is to perform rinse sampling with a specific analytical method for the highly hazardous active (or a swab sample where rinse sampling is not practical).

Okay, now for two editorial comments. One of the proposed advantages of moving to HBELs (such as ADE/PDE values) was that there could be considerable simplification and cost savings. Now, I believe this is true for highly hazardous actives where it is now no longer required that they be made in dedicated equipment. I'm not so sure that advantage has panned out for actives where the therapeutic effect is the main concern for patient safety. For those latter products, the introduction of HBELs has only made cleaning validation more complex and costly. The advice from EMA about what to do for analytical testing for routine cleaning only further complicates things (unnecessarily in my opinion).

My second editorial comment relates to how these Questions #7 and #8 were added to the EMA 2018 Q&A document. There was *nothing* in the 2016 draft document that *in any way* relates to the issues brought up in these two questions, namely routine analytical testing and using visual limits to document that HBEL limits are being achieved. I would have thought that if the EMA wanted to bring up totally new issues, they would have solicited comments on these two questions. [Note: It may be the case that question 5.5 of the ICH Q7 Q&A [ICH, 2015] is the basis for these EMA questions. However, EMA added some constraints that unnecessarily complicated things.]

As it stands, it appears that HBELs are with us to stay. However, my position on their use as limits has not changed. I am still an advocate that for *all actives* companies use the most stringent of (a) an HBEL calculated limit, (b) a dose-based calculated limit, and (c) a limit calculated based on 10 ppm in the next product.

Chapter 13 addresses some final issues in the EMA 2018 Q&A document.

The above chapter is based on a *Cleaning Memo* originally published in July 2018.

REFERENCES

European Medicines Agency, "Questions and Answers on Implementation of Risk Based Prevention of Cross Contamination in Production and 'Guideline on Setting Health Based Exposure Limits for Use in Risk Identification in the Manufacture of Different Medicinal Products in Shared Facilities' (EMA/CHMP/CVMP/SWP/169430/2012)". Document EMA/CHMP/CVMP/246844/2018. 19 April 2018.

ICH, *Q7 Guideline: Good Manufacturing Practice Guide for Active Pharmaceutical Ingredients: Questions and Answers.* ICH Secretariat, Geneva, Switzerland, 2015.

13 Other Issues in EMA's Q&A

This is the third chapter that covers additional issues in EMA's 2018 Q&A on limits in shared facilities [EMA, 2018].

Question #4 (Q4) answers the question as to what *competencies* are needed for people developing an HBEL. In the draft version of this Q&A, the answer for the equivalent question (given in Question #9 in that earlier version) was simply looking at the person's "experience and qualification". In the finalized version, the answer is threefold. The person should have (a) "expertise and experience in toxicology/pharmacology", (b) "familiarity with pharmaceuticals", and (c) experience in the development of HBELs (such as OELs or PDEs). Furthermore, when outside experts are contracted for developing HBELs there should be a contractual agreement consistent with Chapter 7 of the EU GMPs [European Commission, 2012]. That "agreement" in Chapter 7 is one dealing with "Outsourced Activities". Finally, in answer to this Q4, the EMA comments about "purchasing" HBEL monographs, with the statement that if those assessments are purchased there must be a "recording" of the "suitability of the provider (including the specific technical expert) as a qualified contractor". Although not specifically addressed by the EMA, it is probably appropriate for assessments done by in-house toxicologists/pharmacologists that their suitability (that is, based on the threefold qualifications stated above) be confirmed.

Question #5 (Q5) deals with HBELs for contract manufacturers. The answer states that the *contract giver* should either (a) provide a "full HBEL assessment" to the contract manufacturers, or (b) provide data to allow the contract manufacturer to perform that assessment. I assume that the reason in "a" for requiring that the "*full*" assessment (and not just the HBEL value) be provided is that the contract manufacturer is responsible for making sure that the HBEL is appropriately determined since that HBEL is of more concern for determining effects on *other products* (as opposed to effects on the safety of the product the HBEL is associated with) that perhaps are from a *different* contract giver. Furthermore, I would assume that for "b" if the HBEL is developed or contracted out by the CMO, that assessment should be provided by the CMO to the contract giver, because from a GMP perspective, the contract giver is ultimately responsible for cleaning validation of its product.

Question #12 (Q12) deals with HBELs for veterinary products in facilities that make products for different animal species. The first part of the answer is that carryover limits should "generally" be based on the human HBEL. While not specifically stated in the Q&A, consistent with section 4.1 of the 2014 EMA Guideline

[EMA, 2014], the human HBEL should be utilized for veterinary products not as a mass value but rather as a mass/body weight value (then the body weight of the species taking the *next product* can be used); this helps address the large difference in body weight between a cat and a horse. The second part of the answer deals with specific "known susceptibility" for a species; rather than try to summarize it, the full answer given is "However, in cases where there is concern relating to known susceptibility of a particular species (e.g. monensin in horses) the HBEL approach should take into account knowledge of specific animal toxicity when evaluating products manufactured in shared facilities/equipment".

Note that this EMA Q&A document has been essentially adopted by other regulatory bodies [PIC/S, 2020].

The above chapter is based on a *Cleaning Memo* originally published in August 2018.

REFERENCES

European Commission, "EU Guidelines for Good Manufacturing Practice for Medicinal Products for Human and Veterinary Use". Chapter 7: Outsourced Activities. Ref. Ares(2012)778531. 28 June 2012.

European Medicines Agency, "Guideline On Setting Health Based Exposure Limits for Use in Risk Identification in the Manufacture of Different Medicinal Products in Shared Facilities", Document EMA/CHMP/ CVMP/SWP/169430/2012. London. 20 November 2014.

European Medicines Agency, "Questions and Answers on Implementation of Risk Based Prevention of Cross Contamination in Production and 'Guideline on Setting Health Based Exposure Limits for Use in Risk Identification in the Manufacture of Different Medicinal Products in Shared Facilities' (EMA/CHMP/CVMP/SWP/169430/2012)". Document EMA/CHMP/CVMP/246844/2018. 19 April 2018.

PIC/S, "Questions and Answers on Implementation of Risk-Based Prevention of Crosscontamination in Production and 'Guideline on Setting Health-based Exposure Limits for Use in Risk Identification in the Manufacture of Different Medicinal Products in Shared Facilities'" PI-053-1. June 2020.

14 Highly Hazardous Products in Shared Facilities

This chapter deals with the question of manufacturing highly hazardous products *in the same equipment or in the same facility* as products that are not highly hazardous. The obvious answer to that question is "Yes, it certainly can be done from a regulatory compliance perspective". After all, isn't that the whole point of ISPE's Risk MaPP and the EMA's 2014 guideline on limits in shared facilities [EMA, 2014; ISPE, 2017]. The idea behind both those documents is "the dose makes the poison" principle. So, if we can establish that the dose (that is, the *residue limit*) is low enough by a comprehensive toxicological assessment, then with *appropriate* cleaning validation we make a highly hazardous product in the same equipment with products that are not highly hazardous. By doing so, we can clearly avoid dedicated equipment.

The next question is whether that is a good *business* practice. Some may say it is, because it allows pharmaceutical manufacturers to make a product more efficiently, thus saving costs (and depending on what planet you live on, passing those savings on to patients). However, there are other compliance and business reasons that may lead to the opposite conclusion. The first reason is that the risk to patients, and therefore the business risks to a pharmaceutical manufacturer, is much higher if any pharmaceutical product becomes cross-contaminated with a highly hazardous active. So, there are probably some things that I as a manufacturer should consider to minimize that risk.

Certainly having solid cleaning validation is one of those things. But, we all know that things don't always go right; so we want to set up practices and procedures to assure consistency *after* cleaning validation is completed. For example, we may want to increase what we do for routine monitoring for highly hazardous actives as compared to non-highly hazardous actives. That routine monitoring may include rinse or swab sample testing solely for the highly hazardous active (preferably by a specific analytical method). And it may include enhanced visual inspection, either a more extensive visual inspection or, if possible, visual inspection with a documented visual residue limit for the highly hazardous active.

Other practices may include more robust cleaning procedures as compared to what is ordinarily done for non-highly hazardous actives. This may include longer washing times and/or rinsing times, higher concentrations of detergents, and the use of an oxidizing treatment (either before detergent cleaning or after detergent cleaning) to provide a higher assurance of molecular degradation (thus reducing the inherent toxicity of the residue). It may also include just repeating the cleaning process (that is performing it again from "start to finish" as opposed to just extending the wash/rinse times). Some companies have also manufactured a placebo batch (which is then discarded) between batches of different highly hazardous products.

Other practices utilized for highly hazardous actives may be present not to protect patients, but to *protect cleaning operators*. Certainly, personal protective equipment (PPE) may be different, and this includes the extent and type of gowning, type of gloves used, special breathing equipment, and types of eye and/or facial protection. In addition, the use of some PPE items may involve a decontamination step as the cleaning operators leave the cleaning area or leave the manufacturing suite.

Finally, with highly hazardous actives, there may be more concerns about the transfer of residue from one product to another from *indirect product contact surfaces* or from non-product contact surfaces. These may be from floors, walls, outsides of equipment, or even the manufacturing operators' PPE.

With all these concerns, I may be tempted to throw my hands up and just say I don't want to manufacture highly hazardous actives at all. But that is not the point of this chapter. Most of the issues discussed above are applicable to highly hazardous actives even if they are not made in equipment shared with non-highly hazardous actives. For companies that can't (or don't want to) provide those added safeguards, avoiding highly hazardous actives sounds like a good decision. But that still brings us back to the question of manufacturing highly hazardous active and non-highly active in the same equipment. And my main concern is a practical one. If I manufacture both in the same facility and treat both the same, I have a better assurance of making sure things are done correctly.

But if I have certain procedures for highly hazardous actives and a different (typically less stringent) set of procedures for non-highly hazardous actives, I run the risk of the procedures being used incorrectly. Now if I "accidently" use a more stringent procedure for non-highly hazardous products, from a scientific perspective, I may have little concern (although from a CGMP perspective I am *not* following the approved procedure). In the opposite situation of using less stringent procedures for highly hazardous products, clearly I not only have a CGMP issue of using the wrong procedure, but I may face a significant risk of patient safety, product quality, and/or operator safety.

It is for these reasons that I should pursue a more comprehensive risk assessment (including business risks) before I make highly hazardous products and non-highly hazardous products in the same equipment, suite, or facility.

Note that this is not exactly the same issue discussed in EMA's 2018 Q&A document [EMA, 2018]. Question #9 in that document seems to be addressing the question of thinking that if I place all highly hazardous products in one facility, then I don't have to worry about the implementation of health-based exposure limits. EMA's answer is clear; if you make multiple highly hazardous products in a given equipment train, then you should definitely be concerned about getting the active of one highly hazardous product in a different highly hazardous product.

Of course, that concern is mitigated if the active of the first product is present in the second product at a level below the health-based exposure limit, which is the *same* concern where all products are not highly hazardous, where both products are highly hazardous, and where only one of the products is highly hazardous.

My point is *not* that both highly hazardous and non-highly hazardous products should be made in separate equipment, suites, and/or facilities. What I am saying is to pay more attention to *all* the risks if you do so.

The above chapter is based on a *Cleaning Memo* originally published in February 2020.

REFERENCES

European Medicines Agency. "Guideline on Setting Health Based Exposure Limits for Use in Risk Identification in the Manufacture of Different Medicinal Products in Shared Facilities". Document EMA/CHMP/ CVMP/SWP/169430/2012. London. 20 November 2014.

European Medicines Agency, "Questions and Answers on Implementation of Risk Based Prevention of Cross Contamination in Production and 'Guideline on Setting Health Based Exposure Limits for Use in Risk Identification in the Manufacture of Different Medicinal Products in Shared Facilities' (EMA/CHMP/CVMP/SWP/169430/2012)". Document EMA/CHMP/CVMP/246844/2018. 19 April 2018.

ISPE. *Risk-Based Manufacture of Pharmaceutical Products* 2nd Edition, ISPE, North Bethesda, MD, July 2017.

Section III

Limits: General

These next ten chapters deal with limits in the general sense. For more on HBELs, please consult Chapters 7 through 14.

DOI: 10.1201/9781003366003-17

15 EMA vs. ISPE on Cleaning Limits?

You might be mystified at this title. What am I getting at? Well, what I discuss in this chapter is the difference (or conflict?) between the EMA and ISPE on setting limits for cleaning validation. For the EMA "position", I will refer both to the 2014 "Guideline on setting health based exposure limits for use in risk identification in the manufacture of different medicinal products in shared facilities" [EMA, 2014] and the 2018 "questions and answers" on *implementation* [EMA, 2018] of that 2014 guideline. For the ISPE "position", I will primarily refer to the 2017 revision of "Risk-Based Manufacture of Pharmaceutical Products" [ISPE, 2017]. For clarification, when I refer to the EMA position, I am not implying that everyone in the EMA supports this position; and when I refer to the ISPE position, I am not implying that everyone in the ISPE supports that position. Clearly, I am a member of the ISPE and do not fully support its position.

Now, what are the different positions, and what aspect of limits am I referring to? What I'm referring to are adequate and sufficient approaches and rationales for setting limits for actives (APIs) in drug products. Many of you should be aware of the general approach used before 2010. That approach, popularized by Lilly scientists in a 1993 publication, involved setting limits based on 0.001 of a minimum daily dose of the cleaned active in a maximum daily dose of the next drug product. In addition, if that calculation resulted in a concentration of more than 10 ppm in the next product, then 10 ppm was utilized in place of the value calculated by the 0.001 dose criteria. Also, the equipment was required to be visually clean [Fourman and Mullen, 1993].

Furthermore, while not specifically mentioned in the original publication by the Lilly scientists, that approach was applicable to situations where the primary patient safety concern was the *therapeutic effect* of the active. For situations where *other* toxicity effects (such effects as mutagenicity, reproductive hazards, and cytotoxicity) of the active were critical (I typically refer to these actives as "highly hazardous actives"), the approach was not clear except that making the products in dedicated equipment was certainly an option. Another option listed in the PIC/S PI 006-3 document [PIC/S, 2007] was to make those products in shared facilities but with a cleaning validation limit of non-detectability by the *best available* analytical technique (which is somewhat questionable in that, in the absence of setting appropriate limits for the active, the best available analytical technique *may not* adequately protect patients).

In 2010 and then in a 2017 revision, ISPE's Risk-MaPP [ISPE, 2010; ISPE, 2017] presented a newer approach (which had been presented before in various industry forums). Focusing on the 2017 document, Risk-MaPP states that "The only criteria necessary for a robust cleaning process are the health-based, ADE derived limit, a validated analytical method with sensitivity below the acceptance limit that is visually clean". The second criterion of having an analytical method that can actually quantify residues at the acceptance limit is nothing new, but only common sense. Clearly, the third criterion of visually clean was also nothing new. So that the main point in that assertion from Risk-MaPP (and the focus of this chapter) is the element of setting limits *solely* based on an ADE (or PDE) value, that is, a focus only on patient safety from a toxicological perspective.

Now, what is the EMA position? In its 2014 guideline [EMA, 2014], the approach is not so much setting *cleaning validation* limits, but rather to specify methods for setting *health-based exposure limits* (HBELs). Now you might be thinking, aren't "health-based exposure limits" the *same* as "cleaning validation limits"? I'll let the EMA clarify the difference. In that 2014 document, examples of setting HBELss are PDEs and TTC. But that document also states that in certain situations HBELs based on PDEs are *not* "appropriate", such as for biotech manufacture where the protein actives are degraded by cleaning with hot, aqueous alkaline cleaning solutions. Another example given is for actives where the most relevant data is not animal studies but rather human clinical data. Unfortunately, there is no guidance given in that document on how to set HBELs in those two cases.

Then, in 2018, the EMA clarified its position on setting *cleaning validation limits* in its "Questions and Answers on Limits for Shared Facilities" [EMA, 2018]. In that Q&A document, in answer to Question #6 (which is "How can limits for cleaning purposes be established?"), the EMA states "Although the EMA guideline (EMA/CHMP/CVMP/SWP/169430/2012) may be used to justify cleaning limits (as per Introduction paragraph 3), it is not intended to be used to set cleaning limits at the level of the calculated HBEL". What is this saying? As I read it, there appears to be a distinction between an HBEL and a cleaning validation limit. What is that difference? The answer to Question #6 then goes on to talk about "historically used cleaning limits" *for existing products*. It states quite clearly that those limits "should be retained" and that those cleaning limits can be used as *alert limits* to "provide sufficient assurance that excursions above the HBEL will be prevented".

Even though the 2018 document does not state *explicitly* what those historically used cleaning limits are, they can only be what are called the "traditional limits" of 0.001 of a dose and 10 ppm as given in the EMA's 2016 *draft* Q&A document [EMA, 2016]. The EMA clarifies that those historically used cleaning limits should be used for *existing* products. What then about *new* products; certainly HBELs *alone* should be used? Not so fast. The EMA continues in its answer to Question #6 that "A similar process should be adopted when establishing cleaning alert levels for products introduced into a facility *for the first-time*" [emphasis added] Now, what is meant by a "similar process"? The context would seem to make it clear that the use of "historically used limits" is also applicable for new products.

Now, we can argue about the difference between the terms "cleaning limits" and "cleaning alert levels", but my experience has always been that an alert level should be more stringent than a cleaning validation limit. So if I am correct, the cleaning validation limit should never be higher than the HBEL, but in some cases, it may be lower (that is, more stringent). And if "historically used cleaning limits" are still applicable in combination with HBELs, on what basis can the ISPE in its Risk-MaPP document assert that "The only criteria necessary for a robust cleaning process are the health-based, ADE derived limit, a validated analytical method with a sensitivity below the acceptance limit, that is visually clean". Clearly, there is a conflict, or at least a major difference of opinion, between these two approaches to setting limits for pharmaceutical cleaning validation.

Why Risk-MaPP states what it does is inexplicable to me. While the EMA Q&A document may be confusing in its use of "cleaning validation limits" and "alert limits", I believe its approach has a more logical and more scientific basis for setting cleaning validation limits. From a scientific perspective, the EMA approach seems to be based on potential adverse effects on *both* patient safety *and* product quality, and thus is consistent with the approach of the FDA in the 2015 Question #7 of the FDA's "Questions and Answers on Current Good Manufacturing Practices – Equipment" [FDA, 2015], while the ISPE approach seems to focus *only* on patient safety. The bigger question that the pharmaceutical cleaning validation and toxicological communities should address is how did we ever get into this murky situation? If my experience is any guide, that is a question that will never be openly addressed.

The above chapter is based on a *Cleaning Memo* originally published in January 2020.

REFERENCES

European Medicines Agency, "Guideline on Setting Health Based Exposure Limits for Use in Risk Identification in the Manufacture of Different Medicinal Products in Shared Facilities". Document EMA/CHMP/ CVMP/SWP/169430/2012. London. 20 November 2014.

European Medicines Agency, "Questions and Answers on Implementation of Risk Based Prevention of Cross Contamination in Production and 'Guideline on Setting Health Based Exposure Limits for Use in Risk Identification in the Manufacture of Different Medicinal Products in Shared Facilities' (EMA/CHMP/CVMP/SWP/169430/2012)". Document EMA/CHMP/CVMP/463311/2016. 15 December 2016. http://www.cleaningvalidation.com/files/128455599.pdf (accessed May 3, 2021).

European Medicines Agency, "Questions and Answers on Implementation of Risk Based Prevention of Cross Contamination in Production and 'Guideline on Setting Health Based Exposure Limits for Use in Risk Identification in the Manufacture of Different Medicinal Products in Shared Facilities' (EMA/CHMP/CVMP/SWP/169430/2012)". Document EMA/CHMP/CVMP/246844/2018. 19 April 2018.

FDA. "Questions and Answer on Current Good Manufacturing Practices – Equipment". *Question #7*, 8 June 2015. https://www.fda.gov/drugs/guidances-drugs/questions-and-answers-current-good-manufacturing-practices-equipment (accessed May 4, 2021).

Fourman, G.L.; Mullen, M.V. "Determining Cleaning Validation Acceptance Limits for Pharmaceutical Manufacturing Operations". *Pharm. Technol.* 1993, 17, pp. 54–60.

ISPE. *Risk-Based Manufacture of Pharmaceutical Products* (Risk-MaPP), First Edition, ISPE, North Bethesda, MD, September 2010.

ISPE. *Risk-Based Manufacture of Pharmaceutical Products* 2nd Edition, ISPE, North Bethesda, MD, July 2017.

PIC/S. *Recommendations on Validation Master Plan, Installation and Operational Qualification, Non-Sterile Process Validation Cleaning Validation*, PIC/S, Pharmaceutical Inspection Convention Pharmaceutical Inspection Co-Operation Scheme (PIC/S), Document PI 006-3, Geneva, Switzerland, September 15, 2007.

16 Does a High "Margin of Safety" Protect Patients?

A key concept in ISPE's Risk-MaPP is the "Margin of Safety" [ISPE, 2017]. This chapter focuses on that concept *as defined in Risk-MaPP*. When I use the term "Margin of Safety" (with caps) I am referring to the concept as defined in Risk-MaPP. As used in Risk-MaPP, the phrase is sometimes capitalized ("Margin of Safety"), sometimes not in caps ("margin of safety"), and once is called a "True Margin of Safety". When I refer to a "safety margin", I am using it in a generic sense as that is generally understood.

But before I get into discussing the "Margin of Safety", I will provide some background on Risk-MaPP's setting of limits based on ADEs.

According to Risk-MaPP, the only limit needed for cleaning validation (other than the equipment being visually clean) is one based on an ADE. Section 6.3.2.3 of the 2017 revision states that the "only criteria necessary for a robust cleaning process are the health-based, ADE derived limit, a validated analytical method with a sensitivity below the acceptance limit, that is visually clean". Risk-Mapp also states that a limit based on an ADE adequately protects patients, and setting limits on a *more stringent basis* does not provide *additional* patient protection. This is explicit in Section 6.3.2.3, which goes on to say that "The healthbased limit represents a level that is safe for all patient populations and reducing this further does not increase patient safety...."

Now you will notice there is an ellipsis at the end of that last quote from Risk-MaPP, because the sentence goes on to say, "... and, as discussed above, actually lowers the apparent margin of safety" (that is to say, limits more stringent than the ADE-based limit lower the Margin of Safety). I will leave the discussion of "Margin of Safety" for just a minute, and will cover a related concept in Risk-MaPP. Risk-MaPP also recommends that while acceptance limits are based on an ADE, the actual data achieved in a cleaning validation protocol should be "as low as possible below" the limit based on an ADE. It states this in two different sections. The actual statement in 6.3.2.1 is that data should be evaluated "to ensure that any residuals after cleaning are as low as possible below the health-based criteria...." The actual statement in 6.3.2.6 is "It is important that the residue data is as far below the health-based residue level as possible."

Now, if it is true that reducing limits below a limit based on an ADE does not provide additional patient protection, why is there a call for actual data to be as "low as possible" below the ADE-based limit? Well, that brings us to the concept of "Margin of Safety" as defined in Risk-MaPP. One would think that in a document that emphasizes patient safety, any discussion of a safety margin would be

DOI: 10.1201/9781003366003-19

related to additional *patient* protections. But, that is apparently not the case, since as I have already pointed out, there is the clear statement about reducing the limit (that is, making it more stringent) *not* increasing patient safety. However, in Risk-MaPP the "Margin of Safety" is *defined* as the difference between the actual residue data and the calculated limit. So, there appear to be at least two ways to increase the "Margin of Safety". One way is to achieve lower actual residue data, and a second way is to increase the limit. But, if the residue data is to be as low as possible, is there really a *patient protection advantage* by having a greater "Margin of Safety"? My answer is "no". What Risk-MaPP does achieve is an increased safety margin *for the manufacturer* in that the manufacturer is less likely to have failing results if it can set higher acceptance limits. In other words, the increased "Margin of Safety" offers an advantage to the manufacturer in terms of less likelihood of a validation failure, and not necessarily a safety advantage to the patient.

One way to look at this is to suppose I am making a product and there are two ways to set limits. One way gives me an acceptance limit of 10X ppm in a rinse sample, and another way gives me an acceptance limit of 2X ppm in a rinse sample. Let's say my actual residue data was only 0.1X ppm. Is the risk to the patient different depending on which limit I utilize? I think not. But, I am less likely to have a failing result if my limit is set higher. Furthermore, Risk-MaPP explicitly states that manufacturers should not "relax their cleaning processes" just because a higher limit gives a greater "Margin of Safety". In one sense, this is consistent with statements about actual residue data being as low as possible below ADE-based limits.

Of course, Risk-MaPP is assuming that manufacturers will try to achieve limits as "low as possible". While it is probably true that most firms are not going to change cleaning processes that have previously been validated (as long as any limit based on the traditional approach is more stringent than an ADE-based limit), I am not so sure that for new processes this will be the case.

The upshot of this is that Risk-MaPP's "Margin of Safety" does not provide increased patient protection, but merely reduces the *business risk* of a manufacturer in achieving its residue limits in validation protocols. This "Margin of Safety" concept is a relatively strange concept in any discussion focused on patient risk. The only reason it appears to be in the document is to make ADE-based limits more attractive. While the "ADE only" approach may be more attractive to a manufacturer, except for highly hazardous actives (where ADE-based limits are required for non-dedicated manufacturing), it provides no added patient advantage (unless one believes that by lessening the stringency of the cleaning process pharmaceutical companies would lower the selling price of those drugs).

As a final note, Risk-MaPP advocates may accuse me of quoting the document out of context. I invite all to read the document sections that I have quoted to see that my assertions accurately reflect the inconsistencies in the document. I have now been critiquing the Risk-MaPP approach to limits for at least eight years and I still have not seen valid responses to my concerns other than statements to the effect that ADE-based limits are the "science-based" limits. In my view, Risk-MaPP advocates play the "science-based" card as a way to silence critics. And they have been relatively successful because, apart from the recent *draft* EMA Q&A [EMA, 2016] I appear to be the only one who publically points out the problem with the Risk-MaPP approach. For clarification, the EMA does not directly point out the problems with Risk-MaPP; the 2016 *draft* EMA Q&A merely accepts the *traditional* approach as acceptable for actives that are not highly hazardous.

The above chapter is based on a *Cleaning Memo* originally published in February 2018.

REFERENCES

European Medicines Agency, "Questions and Answers on Implementation of Risk Based Prevention of Cross Contamination in Production and 'Guideline on Setting Health Based Exposure Limits for Use in Risk Identification in the Manufacture of Different Medicinal Products in Shared Facilities' (EMA/CHMP/CVMP/SWP/169430/2012)". Document EMA/CHMP/CVMP/463311/2016. 15 December 2016. http://www.cleaningvalidation.com/files/128455599.pdf (accessed May 3, 2021).

ISPE. *Risk-Based Manufacture of Pharmaceutical Products* 2nd Edition, ISPE, North Bethesda, MD, July 2017.

17 What If the *Next* Product Is the *Same* Product?

In performing carryover calculations, we usually talk about limits for the active of Product A if Product B is the next product. Can carryover calculations be done if Product A is cleaned and then Product A is the next product? For clarification, this generally comes up in the context of there being two or more products made on the same equipment. So, this is *not necessarily* a situation of equipment dedicated to one product (although what follows could be applied to that situation). And, it is *not* the situation of cleaning in a campaign where minor cleaning is done between batches of the same product, and where a validated cleaning is only done at the end of that campaign.

So, what can be done? The key is deciding what to do for the L1 calculation, where L1 (the concentration limit in the next product) is the limit in the *next* product. For review, what is typically done for this type of calculation using (for illustration purposes) dose-based limits is as follows. The L1 limit is expressed as 0.001 of a minimum daily dose of the active of the *cleaned* product A divided by the maximum daily drug product dose of the *next* drug product B. This is expressed mathematically as:

$$L1 = (0.001)(MinD_{ActA})/(MaxD_{ProdB}) \qquad (17.1)$$

where
$MinD_{ActA}$ = Minimum daily dose of the active of drug product A
$MaxD_{ProdB}$ = Maximum daily dose of drug product B

So, the question arises, what if the cleaned product is A and the next product is *also* A? Do I follow the same equation with a minimum dose of the active in A in a maximum dose of the product A itself? While it is possible to do so, it is not required logically or from a scientific perspective. Why is that the case? And what is the alternative if that calculation is not required?

If firms want to use the same type of carryover equation for an L1 value for the "A→A" situation, the recommendation I usually have is to require an L1 of 0.001 of the product active concentration. Expressed mathematically, this is:

$$L1 = (0.001)(Conc_{ActA}) \qquad (17.2)$$

DOI: 10.1201/9781003366003-20

where

$Conc_{ActA}$ = Concentration of the active in drug product A

Note that dosing (minimum or maximum) doesn't come into play here. So this type of formulation for A→A could also be used for highly hazardous actives or any situation where ADE/PDE values are used for the typical L1 equation.

For the situation with two *different* products, the reason that we specify the minimum for A and the maximum for B in Equation 17.1 is that we don't know what the appropriate dose for Product A would be for a specific patient taking the maximum dose of Product B. So we use as a worst case the lowest dose of A. But, *if both products are the same*, then it doesn't matter whether the patient is taking the minimum dose or the maximum dose; the 0.001 factor provides the "equivalent" protection in terms of patient safety. But the main concern in this situation is *not* patient safety. Where the cleaned product and the next product are the same, the *real effect* of carrying over the *same* active into the next batch is to change the concentration of the active in that second batch. Of course, there are exceptions, such as when the active degrades during the cleaning process. In that case, it is necessary to consider a toxicity limit based on that specific degradant.

Back to the main case, here is an example of setting limits where the active is not degraded. If the concentration of active in a tablet is 5% (for example, 20 mg active in a 400 mg tablet), then the L1 (the concentration limit in the next product) would be one one-thousandth of that or 50 ppm. So, the second batch containing not 5% active but rather 5.005% active would *not* be a significant patient safety concern or a significant product quality concern. In fact, it probably would be well with the normal potency specification for that product.

Here is an *analogy* that might help with the logic. Let's say I am dealing with two *different* products (A and B) each of which is dosed differently for adults and for children, and where for each product the child dose is *always smaller* than the adult dose. For the cleaning of A, do I need to set limits based on the *minimum child* daily dose of the active in A divided by the *maximum adult* daily dose of drug product B? The answer is I could, and it would be safe in the situation discussed (just for clarification, it would not be if the dose for children was *greater* than the dose for adults). However, it should *not* be required. Why? Because in that situation the relevant safe dose of the active of A in Product B (where product B is taken by an adult) is the *adult* dose the active of A. So in this situation, I recommend that firms consider calculating the child-to-child dose and also the adult-to-adult dose, and use the *more stringent* (lower) of the two L1 values for subsequent calculations. Note that in this analogy, I am not going from A to A, but from A to B, so the analogy is not perfect. Note that if I were using an HBEL calculation in this situation, I might consider a child's HBEL of the active of A in a maximum child dose of product B, and similarly for the adult situation.

Let's get back to the situation of "A to A" limits. For clarification, this formulation of "A → A" cleaning in Equation 17.2 is not necessarily required. It still *may* be appropriate to only require visually clean as a limit for the active (while still measuring bioburden and cleaning agent in the protocol). There is also the possibility that some firms *might* have a rationale to use a less stringent safety factor (such as 0.01). Furthermore, since the context of where Equation 17.2 is usually done is in a situation where I am making multiple products on the same equipment, I still do calculations for "A→A", "A→B", "A→C", and so on to eventually arrive at an L3 value (limit per surface area) for each situation. Then, I compare the different L3 values for the cleaning of A and use the lowest (the most stringent) to set acceptance criteria for my protocol for the validated cleaning of A.

This sounds like a lot on a topic that might not commonly come up. However, it always helps to understand the rationale for what is done if we choose to do it. Chapter 18 also further elaborates on the particular situation.

The above chapter is based on a *Cleaning Memo* originally published in September 2018.

18 Limits for "Product A to Product A"

Typically, the greatest concern in limits for drug actives is when moving from one drug product to a *different* drug product. This is sometimes expressed as the limit "ProdA to ProdB". For this chapter, we will cover the situation where the *same* product follows the cleaned product; that is, it will be a "ProdA to ProdA" situation. In discussing this, we will cover limits set on HBELs as well as limits set on a 0.001 dose criterion. Note in this discussion, we will stop at the L1 limit (the concentration in the next drug product), because beyond L1, there is nothing significant or different in the "A → A" situation as compared to the "A → B" situation. Furthermore, we'll use drug product manufacturing for examples, and not drug substance manufacturing (although the principles apply there as well). Finally, this discussion focuses on situations where there is *validated* cleaning between the two batches of ProdA, whether this is dedicated equipment or consecutive campaigns of the same product.

We'll cover the *dose-based* situation first. The L1 limit for the "A → B" situation is typically calculated based on allowing no more than 0.001 of the *minimum* daily dose of the active in ProdA in a *maximum* daily dose of the *drug product ProdB*. Why is it a matter of *minimums and maximums*? A simple answer might be that it represents a worst case (a lower limit) because the minimum value is in the numerator and the maximum value is in the denominator. But, as I regularly teach, there are some worst cases that must be used, and there are some worst cases that *could* be used, but are *not* required by good science or sound logic. In this example (going from one product to a different product), why *must* the worst case be used?

The rationale is that the carryover concern is based on what will be the effect of the residue of the active in ProdA on the patient taking ProdB if that residue of ProdA is present in ProdB. Clearly, assuming a fixed concentration of the residue of active of ProdA in ProdB, the worst case (the maximum taken on a daily basis) is if that patient takes the *maximum* dose of ProdB. The question then becomes "On what basis is the safe amount of the active of ProdA set?" It may be that some patients who are taking ProdB might, if they were prescribed ProdA at the same time, would be prescribed ProdA at the *minimum* dose of ProdA, while some other patients might be prescribed at the *maximum* dose of ProdA (and perhaps some an in-between dose). However, there is *no way of consistently knowing* which dose of ProdA would be *relevant* (and therefore applicable) for each person taking ProdB. Therefore, a worst-case assumption is made that for patients taking ProdB, the relevant safe amount of the active of ProdA is based on the *minimum*

DOI: 10.1201/9781003366003-21

dose of the active of ProdA. Note that in this situation, patients taking lower daily doses of ProdB will receive lower daily amounts of the active of ProdA (as compared to those patients where ProdB is taken at the maximum daily dose).

Now we'll move from that situation (based on ProdA and ProdB being two *different* products) to the situation where both the cleaned product and the next product are the *same* product (ProdA → ProdA). The question sometimes arises as to whether in this situation the limit should be based on 0.001 of the minimum daily dose of the active in ProdA in a maximum daily dose of the drug product ProdA. I assume the logic behind this is that if the minimum/maximum calculation works for two different products (where it is a required *worst* case), shouldn't it also be applicable if we are going from one product to the *same* next product? I think *not*. In this situation, *we do know* exactly what the relevant dose of the product is for a given patient. Let's say that the possible doses for ProdA are 1, 2, and 3 tablets per day. Suppose a patient takes 3 doses per day of ProdA. If the safe level of the active of ProdA is 0.001 of a dose *for that patient*, then the *relevant* dose of the cleaned product for that particular patient is based on 3 tablets. In that case, both the numerator and denominator are based on a dosage of 3 tablets. At the other extreme, suppose a given patient only takes 1 tablet per day of ProdA. If the safe level of the active of ProdA is 0.001 of a dose, then the *relevant* dose of the cleaned product for that particular patient is based on 1 tablet. And in that case, both the numerator and denominator are based on a dosage of 1 tablet. Therefore, where both products are ProdA, the *use of minimum and maximum is not relevant* to the potential effect on patient safety for the person taking only ProdA.

Another way to look at this situation of both products being the same product is to evaluate the effect of one active getting carried over to a *different batch* of the same product at a level of 0.001 of concentration of the active (which is essentially the consequence of not using the minimum/maximum formulation in the carryover calculation). Let's say a tablet nominally contains 20 mg of active. If an amount of active corresponding to 0.001 is carried over to a next batch, it will increase the amount of drug active in that next batch by 0.02 mg, thus making the total amount of active in that next batch 20.02 mg. This change in active level would ordinarily be considered an insignificant change not affecting the therapeutic effect or the product quality of that drug product.

Okay, let's change the focus and assume limits are set based on an HBEL (such as a PDE or ADE) of the active of the cleaned product. In that case, if we are considering a "ProdA to ProdB" scenario, then the L1 value is the HBEL value of the active of ProdA divided by the maximum daily dose of ProdB. There is no need to consider minimums *and* maximums (only the maximum of ProdB is relevant). The rationale for using the maximum daily dose of ProdB should be fairly obvious.

But, what happens if we are in the "ProdA to ProdA" situation and we are using HBEL values for limits of the active? Do we say nothing changes, and we still use the HBEL value in the maximum daily dose of ProdA? That is possible, but the issue is that I am not adding a *foreign* residue by carryover of the active; I am merely changing the concentration of the active in the subsequent batch.

I would argue that it makes more sense to just use the criterion of 0.001 of active concentration as the L1 value. This provides a consistency in terms of assessing the possible changes of the active's concentration in the next batch. Otherwise, depending on the ratio of the HBEL to the daily dose, there could be cases with much more than a 0.1% concentration change or with much less than a 0.1% concentration change in the level of the active in the next product.

This Cleaning Memo is not to say that this is the only way to set limits for active in a ProdA to ProdA situation. Depending on the specifics of product manufacture and the manufacturing/cleaning process, some companies might just use a "default" L1 limit of 10 ppm, while others might just require that the equipment be visually clean. Furthermore, it should be clear that in a "ProdA to ProdA" scenario, the issue of carryover of cleaning agents and bioburden still requires consideration in just the same manner as would be used in a "ProdA to ProdB" scenario. Finally, if the active degrades or if there are cleaning process degradants, further assessment of cleaning limits is required.

The above chapter is based on a *Cleaning Memo* originally published in October 2019.

19 Surfaces Areas in Carryover Calculations

The *shared* surface area between two products is one of the key factors used in a typical carryover calculation to determine limits for cleaning validation. The value for the shared surface area (typically in square centimeters or square inches) is in the denominator of a carryover calculation; therefore the *larger* the shared surface area, the *lower* the limit. While I generally like to talk about the *shared* surface area, there are several variations which I have seen used which are not the *actual* shared surface area, but could be values *above* the actual shared areas. The fact that those values are above the actual shared area makes the calculated limits lower, thus representing a worst case and therefore acceptable from a compliance perspective.

Before we explain the alternatives, we'll first consider the technically correct case. If the cleaned product (ProdA) and the subsequently manufactured product (ProdB) are made on exactly the same equipment, then life is simple – the shared surface area is the product contact surface area of the equipment train. If *not all equipment is shared*, then the actual shared surface area is the relevant surface area to use. Let's supposed the equipment used for the two products is as given in Table 19.1, with the check marks indicating which equipment is used for which product.

In this situation, the *actual* surface area shared is from the sum of the areas of EquipQ, EquipR, EquipU and EquipV. EquipS and EquipT are *not* shared between the two products, so their surface areas are excluded. You might ask how limits for EquipT are set in the cleaning of ProdA. The answer is simple – are there additional products (for example, ProdC) which share equipment with ProdA? If so, then the limits for EquipT are established based on the shared equipment between ProdA and ProdC. If EquipT is only used for ProdA, then set limits for EquipT as you would for *dedicated* equipment.

Okay, that is the *technically* correct approach. Now we'll consider two other acceptable approaches used by some companies. One approach is just to use the actual surface area of the cleaned product for the calculation of limits. In the case of ProdA and ProdB as shown in Table 19.1, the limit for cleaning of ProdA (the cleaned product) would include the surface area of the entire train of ProdA. That is, it would include the sum of the surface areas of EquipQ, EquipR, EquipT, EquipU, and EquipV. Wouldn't that give a "shared" area *above* the actual shared area between ProdA and ProdB? Yes, it would, and it would result in a calculated limit *below* that of a calculation using the *actual* shared area. Therefore, it would be a worst case and should be acceptable from a compliance perspective. You

DOI: 10.1201/9781003366003-22

TABLE 19.1
Product and Equipment Matrix

	EquipQ	EquipR	EquipS	EquipT	EquipU	EquipV
ProdA	√	√		√	√	√
ProdB	√	√	√		√	√

might then ask, "Why would someone use this approach?" The answer to that is that it simplifies having to determine what equipment is actually shared, particularly for those not using validated software for the calculation of limits.

Now, here is the second option that some companies use. Again I will illustrate this with the two products in Table 19.1. In this approach, the total actual equipment surface area for each product is calculated, and the value used for the surface area in the calculation is the *lower* of the surface areas of the two individual products. For clarification, this is *not the actual surface area shared*. So for ProdA, the actual surface area would be the sum of the areas of EquipQ, EquipR, EquipT, EquipU, and EquipV. And for ProdB, the actual surface area would be the sum of the areas of EquipQ, EquipR, EquipS, EquipU, and EquipV. By using the lowest surface area of either product train, I am assured that the value used will be *no lower than* the actual shared surface area (another way to state this is that the area used for my calculation will be either the same or greater than the actual shared surface area).

I have illustrated this using only two products. However, the principles can be utilized for situations where three, four, or even more products share some (but not all) equipment.

These alternative approaches (alternatives to the use of the *actual* shared surface area) may drive limits lower. Depending on the types of products, this may or may not make cleaning validation more difficult. If limits are driven too low, then it is always possible to fall back to the approach of using the *actual* shared surface area, thus increasing the calculated limits to some extent.

The above chapter is based on a *Cleaning Memo* originally published in April 2019.

20 Carryover Calculation Errors to Avoid

Carryover calculations are easy to execute, particularly with a software program like eResidue [Quascenta, 2021] or with an Excel spreadsheet. The critical element is making sure the calculations are correct, giving you valid results. This chapter explores different ways that the design of that calculation can go wrong.

First, let me give a typical carryover calculation for a limit of what I call L4b, the concentration of an active in an extracted swab sample for drug product manufacture:

$$L4b = \frac{(SDA)(BS)(SA)}{(MDD)(SSA)(SEA)}$$

where

SDA is the safe daily amount of the active (in mass units like micrograms or milligrams)

BS is the batch size of the next product (in mass units, such as grams or kilograms)

SA is the swabbed area (in area units, such as cm^2)

MDD is the maximum daily dose of the next drug product (in mass units, such as grams)

SSA is the shared surface area for the two products (in area units, such as cm^2)

SEA is the solvent extraction amount, in either mass (grams) or volume (mL) units

In this equation, I am assuming the desired output is an L4b concentration in terms of mass/mass (such as micrograms per gram) or mass/volume (such as micrograms per mL).

So, let's start with simple things like making sure the units are correct. That is, when I enter the values, do the units "cancel out" mathematically so that my desired output is in the correct units? If a term is added or omitted to the above equation that causes the output to be just micrograms, then something is wrong. I should try to find out what is missing, or what needs to added, or what needs to be changed. However, if things are not right, don't just add or subtract a term to make the units correct; make sure that the changes you make not only are correct in terms of the units output but also technically and logically correct in terms of giving the *correct* limit value. I prefer to do this exercise with pencil and paper rather than with a spreadsheet, so that I am clearly seeing what is canceling. Note

DOI: 10.1201/9781003366003-23

here that if the equation is designed to report out mass per swab (L4a), then my expected output could be just grams and not a concentration.

Related to this issue is making sure the units "cancel" appropriately. For example, if I were to enter the BS in kilograms and the MDD in grams, I will not get the correct answer even though both are "mass" units. I would need to include a "units" conversion factor in the equation. In this example, a "units" conversion factor could be to convert the kilograms in the numerator to grams, or to convert the grams in the denominator to kilograms. Another example would be making sure the SEA units are correct. If I want a concentration in mcg/g (ppm), then the SEA units should be in grams; if I want a concentration in mcg/mL, then the SEA units should be in milliliters (realizing that for dilute aqueous solutions with a specific gravity of 1.00, mcg/mL can be considered ppm).

We can take this units issue one step further and make sure that, where appropriate, the units refer to the same thing. The main issue here is that the mass units for BS and for MDD both refer to mass units of the *next* drug product. This should make it clear that the batch size to be entered is *not* the batch size of the cleaned product, but the batch size of the *next* drug product made in the cleaned equipment. Furthermore, while the SDA is an amount (mass) of the *active* in the cleaned drug product, the MDD is the maximum dose of the next *drug product* and *not* of the *active* in the next drug product.

Leaving the units behind, another potential error is assuming that if I move from a dose-based limit to an ADE or PDE limit, I still apply a factor such as 0.001 to the ADE/PDE value. The so-called "safety factor" applied to a minimum daily dose of the cleaned active should *not* be used for the ADE/PDE; the ADE/PDE value is a safe daily amount, and there is no need to apply an additional safety factor. For clarification, you might apply a factor for units conversion of 0.001 to an ADE/PDE value expressed in micrograms, so that the SDA is expressed as milligrams, but that is a *different* use of the 0.001 factor.

Still another potential error is not in the carryover equation itself, but in how the SSA is determined. The SSA is the *total (or cumulative) shared surface area* between the two products. It is not correct to perform *separate* calculations for *each* equipment item in an equipment train. Doing separate calculations, by allowing an SDA in each equipment item in a train, essentially allows a multiple (corresponding to the number of individual equipment items in the train) of the SDA by *cumulative* transfer of residue from each equipment item into the next batch.

There may be variations of the basic equation that individual facilities use which might modify how you make sure the calculations are correct. For example, some firms multiply the calculated limit by the sampling recovery factor expressed as a decimal. That change does not basically change any of the items brought up so far. Another option is where companies use "dose units" (for example, number of tablets) to express both the minimum batch size and the maximum daily dose of the *next* product; if you use that option, make sure you carefully define what a "dose unit" is, since it is easy to get confused.

While I earlier said that you can do carryover calculations for equipment items in a train separately, one exception to that is the use of "stratified sampling"

[LeBlanc, 2013a; LeBlanc, 2013b; LeBlanc, 2013c]. In stratified sampling, it is possible to divide the L2 value among the different equipment items (either arbitrarily or based on such factors as the surface area), as long as the *total* carryover is no more than the L2 value.

As in other things in cleaning validation, design is important. So make sure your carryover equations are correct for your situation; otherwise, erroneous limit values will be used in your cleaning validation protocols.

The above chapter is based on a *Cleaning Memo* originally published in March 2018.

REFERENCES

LeBlanc, D.A., "Basics of 'Stratified Sampling'", in *Cleaning Validation: Practical Compliance Solutions for Pharmaceutical Manufacturing, Volume 3*. Parenteral Drug Association, Bethesda, MD, 2013a, pp. 93–96.

LeBlanc, D.A., "More on 'Stratified Sampling'", in *Cleaning Validation: Practical Compliance Solutions for Pharmaceutical Manufacturing, Volume 3*. Parenteral Drug Association, Bethesda, MD, 2013b, pp. 97–100.

LeBlanc, D.A., "Final Notes on 'Stratified Sampling'", in *Cleaning Validation: Practical Compliance Solutions for Pharmaceutical Manufacturing, Volume 3*. Parenteral Drug Association, Bethesda, MD, 2013c, pp. 101–104.

Quascenta Pte. Ltd, "eResidue: Cleaning Validation Simplified". (2021). https://www. eresidue.com/ (accessed May 4, 2021).

21 Protocol Limits for Yeasts/Molds?

Meeting protocol requirements for bioburden is generally not a significant problem provided the final rinse is with *hot* Purified Water (PW) or *hot* Water for Injection (WFI). It is *not* likely that such a water final rinse would contribute to bioburden, and it is likely that the hot temperature would significantly *reduce* any vegetative microorganisms present in the equipment at the end of the cleaning process. This general rule also applies to situations where alcohol (such as 70% IPA) is applied to the equipment surfaces as a final step. Note that if tap water were the final rinse, then that use certainly would be of greater concern. But those issues are not the main focus of this chapter. The focus is whether limits for yeasts/molds should be included in our *testing* for cleaning process protocols.

The rationale for having limits for yeasts/molds is that there are USP <1111> recommended standards for yeasts/molds in pharmaceutical drug products [USP, 2016]. USP <1111> provides maximum values for aerobic bacteria (Total Aerobic Microbial Count, or TAMC) and for yeasts/molds (Total Combined Yeast and Mold Count, or TCYMC) in non-sterile drug products. For each drug product form (such as oral solids, oral liquids, and topicals), there is a consistent *one-log difference* between the recommended value for TAMC and the recommended value for TCYMC. For example, for solid orals (such as tablets), the value for TAMC is 10^3 CFU/gram (1,000 CFU/gram), while the value for TCYMC is 10^2 CFU/gram (100 CFU/gram). If that same ratio for drug products were applied to cleaning validation samples from surfaces, then we should consider applying a factor of 1/10 to the bacterial count to derive an acceptable level for yeasts/molds. Using that analogy, if the swab limit were 50 CFU/swab, then the limit for yeasts/molds would be only 5 CFU/swab. Ideally, this would involve taking two samples from each (adjacent) location and using different media and different incubation temperatures for the two different microbial concerns (bacteria vs. yeasts/molds).

An argument could be made *against* that one-log difference approach. It would go something like this: The typical limit used for cleaning validation purposes is much lower than what would be needed to provide levels in the next manufactured product near or above the <1111> recommended values; therefore it is not necessary to apply the one log reduction to the TAMC to get an acceptable value for TCYMC. I can understand the logic there, but I'm not sure all would fully support such an approach.

Here is an alternative possible approach that might be acceptable and easier to implement, particularly if in my specific facility I consistently get TAMC values below 5 CFU/swab. With such low total counts, I reduce the likelihood of fast-growing bacteria "inhibiting" the colony formation of yeasts/molds. Therefore

DOI: 10.1201/9781003366003-24

(with your company's microbiologist "buying into" this), you could just do a Total Viable Count (including both bacteria and yeasts/molds). For ease of calculation for this example, I am going to use an acceptance criterion of 50 CFU/swab for the total count. Then I set my yeast/mold count at one log smaller, or at 5 CFU/swab. Therefore, if my total count (bacteria and yeasts/molds) were 5 CFU/swab or less, then I can safely assume that I have acceptable levels of *both* bacteria and yeasts/molds. But, if my total counts were 10 CFU/swab, I would characterize the colonies to determine if they were yeasts/molds. This might be something that could be simply done by a visual inspection by a trained microbiologist as to the physical appearance of the colonies, or I might have to do more. In any case, if the number of colonies that were yeasts/molds could be clearly determined to be 5 CFU/swab or less, then I could establish that I was meeting my acceptance limit of 5 CFU/swab. This technique might involve incubating on tryptic soy agar at a temperature intermediate (such as 25°C) [Rhodes et al., 2016]. If acceptable, it simplifies the amount of testing. Note that if I consistently got results of 20 CFU/swab or more, it may not be as helpful. However, if I were to consistently get results above 20 CFU/swab, I should probably look at my cleaning process to see if it could be improved to consistently get values below 5 CFU/swab.

In applying this approach, remember that for cleanroom monitoring, there is generally no distinction in terms of total counts between bacteria and molds/yeasts; it is the *total* combined count that is relevant. Certainly knowing whether it is a bacterium or a yeast/mold could be important for other reasons, but for routine monitoring, it is the total count that is important. Also, remember that if you consistently get values below 5 CFU/swab (with a swab being 25 cm^2), you probably have an acceptable program and you probably have other areas where your efforts could be spent for the continuous improvement of your equipment cleaning validation program.

The above chapter is based on a *Cleaning Memo* originally published in April 2020.

REFERENCES

Rhodes, J; Feasbey, J; Goddard, W; Beaney, A; Baker, M. "The Use of a Single Growth Medium for Environmental Monitoring of Pharmacy Aseptic Units Using Tryptone Soya Agar with 1% Glucose". *European Journal of Parenteral & Pharmaceutical Sciences* 2016, 21(2), pp. 50–55.

USP <1111>, "Microbial Attributes of Nonsterile Pharmaceutical Products". (2016) https://www.usp.org/sites/default/files/usp/document/harmonization/gen-method/q05c_pf_ira_33_2_2007.pdf (accessed May 6, 2021).

22 Cleaning Validation for Homeopathic Drug Products

In order to discuss a possible approach to cleaning validation for homeopathic drugs, I first need to clarify what constitutes a homeopathic drug. "Homeopathy" is a practice in which a substance which causes disease symptoms at high levels is thought to have curative effects of those same symptoms at very low levels [FDA, 2019]. A homeopathic drug is one where the substance is prepared in a more concentrated form, and then diluted down to a very low level. That low level is one where the substance may not be able to be analyzed in the final product. It is for that reason that homeopathic drugs are exempt from the FDA requirement in 211.165 for "laboratory determination of identity and strength of each active ingredient prior to release for distribution". Furthermore, while homeopathic drugs in the USA are not evaluated by the FDA for safety and efficacy, the manufacture of those drugs is subject to the CGMPs [FDA, 2019].

This brings us to the issue of cleaning validation for such products. The conundrum is that if it is *not* practical to have laboratory analysis for the "active" drug in the homeopathic drug itself, on what basis can we determine that the level of the "active" left behind in cleaned equipment does not affect in some way the subsequently manufactured homeopathic product. It probably is not acceptable just to say that at those very low levels it doesn't matter. It also probably is not acceptable to say that adequate further dilution assures removal, because performing recovery studies to demonstrate that fact are also not possible due to analytical limitations. Trying to establish ADE/PDE values probably won't work because it is likely that the ADE/PDE values are already above the levels of the active in the homeopathic product itself.

This brings me to one possible approach (or at least the best approach I could think of). In this approach, I propose something similar to a dose-based calculation that allows 0.001 of the minimum daily dose of cleaned active in the maximum daily dose of the next homeopathic drug. Okay, I know you are probably thinking that that approach will not work for the same reason I gave above, namely that even lower values would not be measurable in any reasonable analytical test. And you would be right!

However, what I generally propose is not to analytically measure the homeopathic active itself, but to measure some other chemical species (which I will call a "marker") in the product. That other chemical species might be an organic salt (where I might measure an ionic species such as sodium ion or phosphate ion), or

DOI: 10.1201/9781003366003-25

it might be the preservative in a formulation (assuming that these products are liquids rather than solids). So, how would that help? Well, I would contend that *if* there is a reasonable expectation that the active would be reduced in the cleaning process by an equivalent percentage as compared to the marker, then analytically measuring the "marker" to show that the marker would be present in the next product at a level of no more than 0.001 of the minimum "dose" of the marker would be acceptable for residues of the active. In general, since the homeopathic drug active is prepared by dilution, it is reasonable that at first pass the reduction by cleaning/rinsing of the product should result in at least an equivalent percent reduction of the drug active and the marker.

That said, I can envision that there may be cases where the solubility of the drug active in the drug product is enhanced by the presence of a marker or another chemical species, resulting in the active possibly not being reduced proportionally to the marker. Another concern might be if the drug product is *dried* on equipment surfaces such that the *dynamics* of removal of the drug active were significantly less than that of the marker. There may be other specific situations that should be evaluated in individual cases.

I should say that I have no idea if this approach would be acceptable to regulatory authorities. However, I suspect that most regulatory authorities would focus on other concerns for homeopathic drugs, even though the October 2019 FDA draft guidance on homeopathic drugs includes "significant deviations from current good manufacturing practice requirements" as one area of focus. The key word is whether not implementing a cleaning validation program would be a *significant* issue as a deviation from CGMPs.

Needless to say, issues related to the control of residues of cleaning agents, microorganisms, and endotoxin are still concerns to be addressed as part of a cleaning validation program for homeopathic drugs.

The above chapter is based on a *Cleaning Memo* originally published in September 2019.

REFERENCES

FDA. "Drug Products Labeled as Homeopathic - Guidance for FDA Staff and Industry". Revised October 2019. https://www.fda.gov/media/131978/download (accessed May 5, 2021).

23 A Possible Approach for Biotech Limits

For years, the biotech industry has argued that cleaning validation limits for biotech manufacture should not be based on the safety of the native protein, because those proteins are deactivated and degraded during a cleaning process with hot, alkaline aqueous cleaning solutions [PDA, 2013]. The 2014 EMA guide on limits for "shared facilities" finally provides some regulatory support for this assertion [EMA, 2014]. That guide (Section 5.3) states:

> Therapeutic macromolecules and peptides are known to degrade and denature when exposed to pH extremes and/or heat, and may become pharmacologically inactive. The cleaning of biopharmaceutical manufacturing equipment is typically performed under conditions which expose equipment surfaces to pH extremes and/or heat, which would lead to the degradation and inactivation of protein-based products. In view of this, the determination of health based exposure limits using PDE limits of the active and intact product may not be required.

This is great, but if PDE limits based on the intact active may not be required, what is required (or what is suggested or recommended)? Nothing in that section of the EMA document clarifies this. It has been suggested in the past [PDA, 2013] that if health-based limits could be established for the degraded fragments, that may be one way to deal more scientifically with biotech limits. The question has then been how do we determine an ADE or PDE value for the degraded protein fragments?

That said, it does appear that there may be help in Section 5.5 of that EMA guidance. That section deals with "Investigational Medicinal Products" (and not specifically biotech). It states:

> For early development (Phase I/II) investigational medicinal products (IMPs) estimation of PDEs may be difficult based on their limited data sets. Where this is apparent, an alternative approach using categorisation into specific default value categories e.g. based on low/high expected pharmacological potency, low/high toxicity, genotoxicity/carcinogenicity, similar to the tiered Threshold of Toxicological Concern approaches proposed by Kroes et al. (2004), Munro et al. (2008), and Dolan et al. (2005)[2], can be considered to derive health-based exposure limits if adequately justified.

Note that I have left out the bibliographic references. However, one of the references is by scientists at multiple large pharma companies [Dolan et al., 2005].

DOI: 10.1201/9781003366003-26

Three of the authors were all with Merck at that time of publication. Dolan is now with Amgen; Naumann was one of the leaders in ISPE's Risk-MaPP. (I'm just trying to make it clear that this was a publication by recognized pharmaceutical toxicologists.)

The basic argument by Dolan et al. is as follows. For "relatively unstudied compounds" with "limited or no toxicity data", an approach similar to the TTC (threshold of toxicological concern) used for genotoxic materials may be used. The tiered approach of safe daily amounts (called ADI values in the publication) is as follows:

Compounds that may be carcinogenic: 1 μg/day
Compounds that may be potent or highly toxic: 10 μg/day
Compounds likely to be none of the above: 100 μg/day

What follows is my suggestion based on what is in the EMA guidance and what is proposed in the Dolan et al publication. (I want to make it clear that this suggestion is *not* put forth by the EMA or by Dolan et al., so I don't know if either would accept it.)

If limits for biotech are not set on the PDE of the active protein, then can it be set on the PDE of the degraded protein? If so, what data could be supportive? We already know that in general the immunotoxicity of proteins is lessened as the protein molecular weight decreases [FDA, 2002; FDA, 2014]. We can also assume that some of these degradants are present as the protein actives degrade in the human body after administration. But what other hard data is available? It would appear that degradants of biotech active proteins would fall under the Dolan et al. category of compounds with limited or no toxicity data. This is a question that has to be determined by toxicologists and pharmacologists of biotech companies. If the answer is that degraded protein actives fall under that category, then the second question is which of the three tiers is appropriate? If they can come to the conclusion that the degraded proteins are not likely to be carcinogenic and not likely to be potent or highly toxic, then it seems reasonable to use the Dolan et al. tiered approach and establish a PDE value of 100 μg/day. I suspect that initially, this is a decision that each biotech company should make for their specific protein actives based on deactivation and degradation data for those specific protein actives.

If that value of 100 μg/day were used for carryover calculations (either for the finished drug manufacture or for equipment after the last purification step in bulk active manufacture) and then expressed as TOC, it is likely that the result would be TOC limits above typical values now used by biotech manufacturers, such as 5–10 ppm TOC for bulk active manufacture and 1–2 ppm TOC for finished drug product manufacture [Parenteral Drug Association, 2013]. In other words, an appeal to the EMA guideline and the Dolan et al. approach (cited in the EMA guideline), along with an assessment of where the degraded fragments fit in the tiered approach, would provide a more scientific rationale for claiming that the current TOC limits are acceptable from a patient safety perspective.

I would not recommend (at least at this time) that any firm increase its TOC limits based on this type of assessment. One of the reasons for this is that (as I have tried to argue many times) the effects of residues that we address in cleaning validation should not be based on patient safety alone; we should also consider effects on product quality (including stability, physical properties, and bioavailability of the active), which may cause the limits to be more stringent. Furthermore, for the early stages of biotech manufacture (fermentation and cell culture), I may be more concerned about the effects of residues on production efficiency and product purity due to interferences with those critical manufacturing processes.

Note that there is another publication by industry scientists proposing a standardized limit for *degraded* protein molecules [Sharnez et al., 2013]. That publication proposes a safe daily amount limit of 650 µg/dose (although this limit was not specifically called an ADE or PDE by the authors, it is possible that this could be considered as a PDE of 650 µg/day for degraded protein actives). That value is based on using "dosing" of a *model* compound, gelatin, to establish a safe dose amount for degraded protein fragments.

Now you might be thinking that there is a big difference between 100 µg/day and 650 µg/day, and see this as a problem. My response would be the Dolan et al. approach is a "one size fits all" approach that is *not* limited to compounds that might be present in biotech manufacture. The Sharnez et al. approach narrows the result based specifically on degraded proteins that might be present in *biotech* manufacture. It is not unlike the fact that the 0.001 dose criterion is a one size fits all approach for non-highly hazardous actives; with a PDE/ADE value for a specific non-highly hazardous active, the safe daily amount value will typically be higher than that determined by the dose criterion.

If this difference between 100 µg/day and 650 µg/day is of concern, consider calculating carryover limits with both values. If both give TOC values above what you currently utilize (which is what I expect in most cases), then stop and just use the lower value (100 µg/day).

If this approach is used, there is one significant consequence for carryover limits. That is, carryover limits will *not* depend on the dosing of the cleaned product. For example, suppose I have one biotech active that is dosed at 1 mg of active per day and a second product that is dosed at 50 mg of active per day. If the PDE value of degraded actives is 100 µg/day, that value will be used in the numerator of a carryover equation in *each* instance. While the dose of the active is irrelevant for the cleaned product, the dose of the drug product or bulk drug active of *each* product will be relevant since it is used in the denominator of the carryover equation *as appropriate for the next product*.

Clearly for this approach to be used, it becomes even more important to establish that the active protein is deactivated and/or degraded in the cleaning process. There may be different ways to design lab studies that could be done to address deactivation and degradation. It may also be possible to actually measure deactivation and degradation in cleaning in commercial manufacture. Finally, additional support for deactivation can be addressed by demonstrating deactivation during any SIP process for equipment.

The purpose of this chapter is not to say we should change our limits for biotech cleaning validation. The purpose is to help provide an even better rationale for the acceptability of current limits.

The above chapter is based on a *Cleaning Memo* originally published in February 2017.

REFERENCES

Dolan, D.G.; Naumann, B.D.; Sargent, E.V.; Maier, A.; Dourson, M. "Application of the Threshold of Toxicological Concern Concept to Pharmaceutical Manufacturing Operations". *Regul. Toxicol. Pharmacol.* 2005, 43, pp. 1–9.

European Medicines Agency, "Guideline on Setting Health Based Exposure Limits for Use in Risk Identification in the Manufacture of Different Medicinal Products in Shared Facilities". Document EMA/CHMP/ CVMP/SWP/169430/2012. London. 20 November 2014.

FDA, "Guidance for Industry: Immunotoxicology Evaluation of Investigational New Drugs". October 2002. https://www.fda.gov/media/72228/download (accessed May 5, 2021).

FDA, "Guidance for Industry: Immunogenicity Assessment for Therapeutic Protein Products". August 2014. https://www.fda.gov/media/85017/download (accessed May 5, 2021).

Parenteral Drug Association, Technical Report No. 49, "Points to Consider for Biotechnology Cleaning Validation", Bethesda, MD, 2013.

Sharnez, R.; Spencer, A.; To, A.; Tholudur, A.; Mytych, D.; Bussiere, J. "Biopharmaceutical Cleaning Validation: Acceptance Limits for Inactivated Product Based on Gelatin as a Reference Impurity". *Jour. Val. Tech.* 2013, 19(1), pp. 1–8.

24 Establishing Clearance for Degraded Protein Actives

In Chapter 23, we covered a possible approach for setting "health-based" limits for biotech manufacture based on the "fact" that the biotech active protein would be degraded and/or inactivated during cleaning. There is an important difference between "inactivation" (or deactivation) and "degradation". Inactivation means that the protein loses specific biological activity based on a change in the protein structure. Degradation usually means that the protein is broken down into smaller molecular weight fragments. In many cases, inactivation and degradation go hand in hand. However, it may be possible to inactivate a protein active without having a significant reduction in the protein's molecular weight. It may also be possible to have degradation without loss of biological activity.

Why am I bringing this up if the issue here is *establishing clearance*? The reason is that biotech firms also expect that in *bulk drug active manufacture*, degraded proteins which may be residues of previous cleaning processes are removed by the various purification processes used in the downstream bulk active manufacture. This is one basis for trying to limit shared surface areas in carryover calculations to only the equipment *after* the last purification process. This approach is based on Section 12.70 of ICH Q7 [ICH, 2000], which states:

> In general, cleaning validation should be directed to situations or process steps where contamination or carryover of materials poses the greatest risk to API quality. For example, in early production it may be unnecessary to validate equipment cleaning procedures where residues are removed by subsequent purification steps.

Purification processes in biotech may involve filtration (for example, where the protein active is retained by the filter based on size or molecular weight), or it may involve chromatography, where the protein active is retained on a column resin (and other components are not), followed by removal or elution of the protein active from the column resin. My preference in establishing that the cleaning process residues are cleared by my downstream processing is to show multiple different ways in which clearance occurs; this is not unlike the approach for viral clearance in showing *orthogonal* processes for virus removal.

In a previous volume in this series [LeBlanc, 2017a; LeBlanc, 2017b; LeBlanc, 2017c] I addressed clearance in *small molecule API* manufacture. In small molecule API organic synthesis, the main clearance mechanisms are filtration and recrystallization, and the focus of that previous publication was both on possible theoretical considerations and on possible studies to demonstrate clearance. This chapter attempts the same for demonstrating clearance of *biotech* cleaning residues. In addressing the biotech situation, we must remember that the purification processes are primarily designed to remove "impurities" from the manufacturing process of a given protein active. Clearance of those impurities is generally addressed in process development and process validation. Clearance of residues left from a *prior cleaning process* is a different matter, particularly if those residues are from a *different* protein active, and even more so if different purification steps are used for the two products (the cleaned product leaving the residues and the next product which could be potentially contaminated). A critical element for residues is not necessarily whether they are removed by the purification steps of the same product (as likely in a biotech campaign of the same product), but whether they are removed by the purification steps of the next product (as likely in any changeover from one product to another).

So, the first step might be taking a look at "theoretical" concerns (theoretical in that these are likely to occur based on sound scientific principles). For example, for filtration (such as tangential flow filtration, or TFF), it might be possible to determine that the active protein being manufactured is retained by a filter based on a molecular weight cutoff of X kDa (kiloDaltons). The molecular weight of the active is >X kDa, so it is retained. I then use data from my degradation study (for example, using SDS-PAGE) on the estimated molecular weight ranges of the degraded fragments from the prior manufactured protein. If the molecular weights of the degraded fragments of the prior manufactured product were found to be <0.5X kDa (I have chosen that 0.5 factor somewhat arbitrarily), then I might conclude that the prior cleaning residues would be cleared from the process stream by that purification step.

The question then becomes "what percentage clearance occurs with that step?" This can be estimated if I know the maximum volume of liquid (or suspension) passed through the filter, and the volume of "liquid" retained by (that is, not passed through) the filter. This assumes that there is no interaction between the protein active and the degraded fragments. One of the functions of this type of filtration, in addition to removing of impurities, is to concentrate the protein. Therefore, you can expect significant clearance of smaller fragments that pass through the filter.

This assessment of clearance may be easily done by this type of "paper" exercise. Actually measuring residue by a technique such as TOC is probably not feasible (there are just too many *other* sources of TOC). That said, it *might* be possible to make an assessment of clearance by performing a "before and after" SDS-PAGE to show that the presence of small molecular weight fragments in the "before" filtration suspension *and* in the filtrate, and the relative lack of such low molecular weight fragments for the retained product. If done on the lab scale, this could also be done using just a preparation of "degraded protein" (without the addition of the next native protein that might be present). This allows not only for SDS-PAGE analysis but also for analysis by TOC and/or total protein.

What about showing clearance in a chromatographic purification process? Because of the variety of chromatographic options (ion exchange, affinity, size exclusion, etc.), it's a little more difficult to make suggestions here. Let's assume that the mechanism is passing a suspension through a column, retention of the protein active by the column (with various *other impurities*, including impurities from the prior cleaning process residues, *flowing through and exiting* the column), and final elution of the protein from the column using a suitable buffer. A paper exercise will be much more difficult, in that predicting retention by the resin may be more difficult.

In such a situation, it may be best to perform a lab study, where the degraded protein alone is passed through the column (simulating actual production conditions). A comparison of the initial degraded protein and the exiting solution by SDS-PAGE, TOC, and total protein may assist in determining clearance by the chromatographic procedure.

Note that in evaluating clearance by filtration and clearance by chromatography, the degraded protein preparation used for lab studies may be different in cases where filtration precedes chromatographic purification.

Finally, the purpose of this chapter is not to say we should change our limits for cleaning validation in biotech bulk drug substance manufacture. The purpose is to help establish clearance and thus help provide an even better rationale for the acceptability of current limits.

The above chapter is based on a *Cleaning Memo* originally published in March 2017.

REFERENCES

ICH Q7, "Good Manufacturing Practice Guide For Active Pharmaceutical Ingredients". 10 November 2000.

LeBlanc, D.A., "Limits for Small Molecule API Synthesis - Part 1", in *Cleaning Validation: Practical Compliance Solutions for Pharmaceutical Manufacturing, Volume 4.* Parenteral Drug Association, Bethesda, MD, 2017a, pp. 71–75.

LeBlanc, D.A., "Limits for Small Molecule API Synthesis - Part 2", in *Cleaning Validation: Practical Compliance Solutions for Pharmaceutical Manufacturing, Volume 4.* Parenteral Drug Association, Bethesda, MD, 2017b, pp. 77–81.

LeBlanc, D.A., "Limits for Small Molecule API Synthesis - Part 3", in *Cleaning Validation: Practical Compliance Solutions for Pharmaceutical Manufacturing, Volume 4.* Parenteral Drug Association, Bethesda, MD, 2017c, pp. 83–85.

Section IV

Visually Clean

The following four chapters deal with issues related to a visually clean evaluation.

DOI: 10.1201/9781003366003-28

2 5 Avoiding "Visually Dirty" Observations

A frequent question I get is "What is the most common FDA (or another regulatory body) finding in an inspection?" Well, I certainly haven't seen them all, and don't have any clear data. However, based on my experience one of the most frequent observations I see is that in an inspection some item of equipment (which could be "product contact" or "non-product contact") is *visually soiled*. Why is that the case? It might be that such an observation is an "easy" one. If the equipment *should be* visually clean, and if it *is not*, that is an easy finding. After all, it is not like seeing the calculations that are done for limits of an active ingredient, where there are all sorts of terms in an equation, and where every company (while using the same principles) expresses those limits in a slightly different way. It takes a lot more effort to see problems in such calculations as compared to looking at equipment and saying "It's not visually clean".

In the chapter, we will look *first* at what we can do to *avoid* situations where equipment is visually soiled. And after that, we will discuss what should be done *in response*, either in an inspection or at any other time, if we find ourselves in a situation where we have visually soiled equipment that *should be visually clean*.

The obvious answer for avoiding visually soiled equipment is to make sure it has been cleaned appropriately, and then that it has been examined and documented that it is in fact visually clean. This means that we have *designed* our validated cleaning procedures (and other cleaning procedures too) well, including the frequency of use of the cleaning procedure. By this I mean it may not be an expectation *within a campaign* (where batch-to-batch "minor" cleaning is done) that the equipment be visually clean. But, for cleaning at the end of a campaign visually clean equipment is usually an expectation. Furthermore, to avoid situations where equipment is documented to be visually clean but where it clearly is *not* visually clean during an inspection, it is critical that those operators doing the visual inspection be appropriately trained. This includes not only *what to look for*, but also *how to look* (with flashlights/torches, at what distance, and the like), *how to document* the visual examination, and *what to do (or who to tell) if the equipment is not visually clean*.

Clearly, there may be cases where the equipment was visually clean at the end of cleaning, and where the visually "clean state" was appropriately documented, but where the "visually soiled" condition occurred *after* the cleaning and *after* a visual examination. We'll discuss how to investigate any visually dirty situation later. However, the "take home" lesson here should be what to do *before* a planned inspection. Certainly, we should prepare so that relevant documents that might be

DOI: 10.1201/9781003366003-29

called for by the inspectors are readily accessible, as well as perhaps review them to make sure they are "inspectable" and "understandable" by a third party. At the same time, it would be prudent to have a *pre-inspection walk-through* to identify any items that might need attention from a "visually clean" perspective. This could include dirty equipment, improperly stored cleaning tools, and/or inaccurate "status" labels. Those would be easy citations. If there is equipment that is "permanently stained" with colors or surface anomalies, be prepared *before* the inspection itself to address those, preferably with an appropriate rationale or justification. Of course, if things are done 100% right in the first place, then there would be no such problems. If you do find such situations, certainly *make corrections*, but also *take corrective actions* to prevent reoccurrences.

Either in a pre-inspection situation or during an inspection (or for that matter at any time), what should be done if I find equipment that *should be* visually clean but clearly *is not*? Avoid the temptation to *immediately* clean it. What I would prefer to do is an investigation to determine the nature of the residue (or soil) on the surface. Determining the type of residue can not only help me determine how to clean it as part of my correction, but also can help me determine any possible product safety/quality impact if there is a possibility that the residue could have transferred to a manufactured product. In addition to determining the nature of the residue, I want to investigate the *root cause* of why it is there (assuming that with proper cleaning, it should *not* be there). Was it operator technique in a manual cleaning process? Was it a problem with the tools used for cleaning? A problem with the cleaning agent used? A problem with the temperature of cleaning? Ideally, a root cause should be identified, and corrective/preventive actions should be identified and implemented. This might include a change in the cleaning parameters. For manual cleaning, this might also include changes in the written procedures to better describe what should be done, as well as retraining of operators.

A key initial question is whether the soil is present throughout the equipment, or only in certain locations. Those certain locations may help you identify the root cause. Furthermore, whether the soiled areas have distinctive, sharp edges suggestive of improper manual scrubbing can be important. The identification of the type of soil (or residue) may include its *physical* nature, its *chemical* nature, and its *microbial* nature. For a physical examination, is the soil particulate or present as a continuous "film"? What color is it? Is it transparent or opaque? Is it highly adherent or loosely adherent to the surface? This may require scraping the surface with a metal or plastic spatula. As you do a physical examination, consider taking samples for chemical analysis. Chemical analysis may include Infrared, HPLC, and UV. As you prepare samples for analysis, observing the solubility of the material in different solvents (aqueous and organic) may be beneficial. Although microorganisms are not likely to be the problem (if you have a visible microbial mass, you *really* have a problem!!), residues may also be characterized for bioburden and/or endotoxin. These characterizations may help you identify not only the correction needed to get the equipment back to its baseline cleanliness, but also what caused the problem. In an investigation, an evaluation of whether this visually soiled condition is a *recurring* problem should be part of looking for a root cause.

Realize that there may be a tension between those who want to quickly do a correction (so that the equipment can be safely used again for manufacturing) and those wanting to do the investigation (to *clearly* identify the root cause so that corrective/preventive actions can be taken). As with many things in life, a balance (and patient understanding) is helpful.

If the problem is seen before subsequent use of the equipment, then provided appropriate correction is done, the impact on a *subsequently* manufactured product is minimal (other than perhaps delaying the production schedule). If the problem is only identified after a *subsequent* product has been made (which should be less likely, but clearly not impossible, on product contact surfaces), then the impact on subsequently produced products should be explored. This may involve estimating the total amount of soil (residue) that may have been left in the equipment. Then, the levels of residues in the next product can be estimated assuming that the residue was *uniformly* mixed in the next manufactured product. That estimated level is then compared to an acceptance limit in the product itself. If the residue could transfer but be *non-uniformly mixed* in the next product, the situation is different, but some adjustments can be made to estimate the impact. This exploration may also involve analysis of the final manufactured product for the suspect residue in that product. This can be more of a challenge, because of the need to have an analytical method that can measure the specific residue in the presence of large amounts of excipients and actives in that *next* product.

Thinking about all these issues *in advance* and preparing approaches to dealing with "visually dirty" equipment (if it were to occur) is a prudent consideration.

The above chapter is based on a *Cleaning Memo* originally published in September 2020.

26 What's a Visual Limit?

Visual Limit or "VL" (also called Visual Residue Limit, or "VRL") is a term dealing with amounts of residue which could be on a surface that are detectable (or not detectable) by visual examination. We all know that the visual detectability of a residue on a surface will depend on such factors as the viewing distance, the lighting, the angle of viewing, the contrast between the residue and surface, and the eyesight of the person who is the viewer. However, those parameters will *not* be the focus of this chapter. Instead, we will focus on how to *define* and how to *determine* the VL. For clarification, this approach is typically used when we want to know whether the VL is a more stringent assessment (as compared to a *calculated* carryover limit) to determine whether the surface is acceptably clean. In other words, the VL is used to determine whether a visually clean surface has residue *below the calculated carryover limit.*

There appear to be at least two general approaches to defining the VL. Both are related to spiking studies (also called spotting studies) in which fixed amounts of residue are placed on a surface, typically as a solvent solution. The spiking is done over a certain surface area (typically a round area based on the solvent spreading across the surface to form a circular shape). Based on the amount of residue spiked and the surface area it is applied to, we are able to determine the concentration of the residue on the surface in terms of mass per area, such as micrograms per square centimeter (mcg/cm^2). Upon evaporation of the solvent, the spiked surface can then be viewed.

In one approach the VL is the *lowest* spiked level in which *any* amount of residue is seen on the spiked surface. That is, at high spiked levels the residue may be visible across the *entire* spiked surface. Then, at lower spiked levels there may be residue visible only on *certain portions* of the spiked area. Eventually, we get to a point where there is *no residue at all* seen on the spiked surface (that is, the surface is visually clean under the defined viewing conditions). The lowest spiked level where there is at least some residue visible is considered the VL.

Now with this approach, you might *first* ask why isn't the highest level at which the surface is *completely* visually clean the VL. The answer to that question is based on the objective of this spiking study. Remember that the objective is to say that a visually clean surface in a protocol has a residue level *below* the calculated carryover limit. Let's suppose I only spike at four levels of 0.5, 1.0, 1.5, and 2.0 mcg/cm^2, and that 0.5 mcg/cm^2 is visually clean and 1.0 mcg/cm^2 is the lowest level at which I can see *any* residue on the spiked surface. Depending on the calculated carryover limit, I can't necessarily say that a VL of 0.5 mcg/cm^2 is appropriate. For example, suppose my calculated limit is 0.8 mcg/cm^2 (note that I have deliberately chosen this limit to illustrate the point.) In that case, if I had spiked at 0.75 mcg/cm^2, I might have seen some residue on the spiked surface, and I could

DOI: 10.1201/9781003366003-30

not be assured that a VL of 0.5 mcg/cm^2 (as defined based on the highest level which is *completely* visually clean), means that I am assured of being below a calculated limit of 0.8 mcg/cm^2.

Now, you might also ask why I would only see residue on a *portion* of the spiked surfaced. The likely reason is that as the solvent evaporates, it does not *evenly* evaporate at the same rate across the entire spiked surface. For example, evaporation may be faster at the edges as opposed to the center of the spiked area. This results in unevenness of dried residue across the surface. In any case, my concern with this approach is related to this issue of uneven appearance of residues on the dried surface. If I spike at a level of 1.0 mcg/cm^2, and see some residue on the surface, does this mean that if 1.0 mcg/cm^2 were spread *evenly* across the surface, that I could *clearly* see it as visible residue? My answer is "No"; the reason is that the small amount of visible residue on the spiked surface probably represents a value of residue in that specific location (and not across the entire spiked surface) of a value *higher* than 1.0 mcg/cm^2. This is due to uneven evaporation, which will cause parts of the surface to be of values *higher* than 1.0 mcg/cm^2 and other parts to be *lower* than 1.0 mcg/cm^2.

This brings me to a second approach for defining the VL (and this is the one I prefer). In that approach, it is still possible to spike coupons at different levels. However, the VL is the *lowest* spiked level where the residue is seen across the *entire* spiked surface. Using the same spiking levels given in the previous example, let's suppose that 0.5 mcg/cm^2 is visually clean, 1.0 mcg/cm^2 is the lowest level at which I can see any residue, and 1.5 mcg/cm^2 is the lowest level where I can see residue across the entire spiked surface. In this last case of a spiked level of 1.5 mcg/cm^2, I suspect that some portions of the spiked surface are slightly above a value of 1.5 mcg/cm^2 and other portions might be slightly below a value of 1.5 mcg/cm^2. But in either case, I would have confidence that any surface observed (again under the same viewing conditions) to be visually clean would have residue values below 1.5 mcg/cm^2. So, as long as my calculated carryover limit was 1.5 mcg/cm^2 (or higher), I could readily use a visual examination (again considering equivalent viewing conditions) to determine that the target residue was below that carryover value. Note further that I am not stating the exact actual value of residue on the surface visually clean surface; I am merely using a visually clean assessment to determine (essentially in a pass/fail test) that I am meeting (that is, less than) my calculated carryover limit.

I typically like to take this one step further. Rather than spiking at multiple levels, a simpler approach is just to spike *at the carryover limit*. For example, if my carryover limit is 2.0 mcg/cm^2, I spike *only* at that level of 2.0 mcg/cm^2. I suspect that because of the unevenness of evaporation, that some portions of the spiked surface will be at levels below 2.0 mcg/cm^2 and that other portions will be above 2.0 mcg/cm^2. But, if I can see residue across the entire spiked surface, then I can have the assurance that any surface viewed as visually clean (again considering equivalent viewing conditions) would have target residue values below 2.0 mcg/cm^2. How much below I can't say, since this is a pass/fail test.

That said, to show the robustness of my visual assessment, for this latter approach, I might spike at a level somewhat below my calculated limit (such as 20% or 40% below) to show the "robustness" of such a visual assessment. The robustness of the assessment might also be addressed by having the viewing conditions in the spiking study to be more "stringent" than the conditions of visual observation in the plant operations. By more "stringent", I mean under conditions that are more likely to result in *higher* values for the VL. For example, lower light levels and longer distances will result in higher VL values. This clarification of the meaning of "more stringent" is necessary because *lower* light levels and *longer* distances are more stringent in the lab study involving spiking studies to determine the VL, but *brighter* light levels and *shorter* distances are more stringent in a protocol evaluation for plant equipment. In the ideal world, the viewing conditions in spiking studies would be exactly the same as conditions on the factory floor, but we don't live in an ideal world. Perhaps the closest you can get to having the viewing conditions exactly the same is to place the spiked coupons in the equipment itself, but this has other concerns such that it may not be practical.

The above chapter is based on a *Cleaning Memo* originally published in August 2019.

27 Visual Residue Limits
Part 1

This chapter supplements the prior chapter in looking at *how to establish visual residue limits* (VRLs). Before I start, here is a little history. Requiring that equipment be visually clean at the beginning of product manufacture is ancient history. The idea that we should require the equipment to be visually clean at the end of a validated cleaning process is a logical extension of that older idea. If the equipment needs to be visually clean before I start to manufacture, it probably must also be visually clean at the end of the cleaning of the prior product; if it is visually soiled at the end of that cleaning process, in most cases, it will not clean itself before I start manufacture of the next product. That the equipment in a validated cleaning process should be visually clean was more or less "solidified" by the Fourman/Mullen 1993 publication, by the FDA Cleaning Validation Guidance of 1993, and by the 2001 PIC/S Cleaning Validation Recommendations [FDA, 1993; Fourman and Mullen, 1993; PIC/S, 2001]. In 2002, I published a paper (with an admittedly *faulty* conclusion) dealing with spiking studies to establish a VRL [LeBlanc, 2002]. There have been many papers published since that time by scientists such as Richard Forsyth (first when he was with Merck & Co., and later with other affiliations) [Forsyth, 2004; Forsyth, 2005]. There are also many chapters in previous volumes of this series dealing with "visually clean", but *not* all deal with VRLs [LeBlanc, 2006; LeBlanc, 2013a; LeBlanc, 2013b; LeBlanc, 2017].

The general approach in a spiking study is to apply a residue at *different levels* on a surface. The spiking is done by dissolving the residue in a volatile solvent, spiking the solution on a coupon of a specified material of construction (MOC), and then drying the spiked coupon under defined conditions. While different companies may specify the spiking levels by the concentration of the residue in the spiking solution or by the concentration of the residue in the extracted swab solution (assuming the calculated limit is for swab sampling), my preference is to define the spiked levels as the *amount per surface area* (such as mcg/cm^2), which corresponds to what I typically call an L3 limit. The spiked, dried coupons are then evaluated under specified conditions of lighting and distance. The evaluation is typically done by several (minimum of three) observers *independently*.

Now we get to the interesting part: how the endpoint of the evaluation is defined. Realize that if we spike at a series of levels, we might see some that are *grossly soiled* over the entire spiked area, some that are *lightly soiled* over the entire spiked area, some that are soiled over only *portions* of the spiked area (that is, some portions are soiled and some portions are visually clean), and some that are *visually clean over the entire spiked area*.

DOI: 10.1201/9781003366003-31

So the question arises: How do I define the VRL? The answer to that depends on how we are using the VRL. The most *useful* purpose is to say that the VRL establishes a value (let's call that value X mcg/cm^2) so that if the equipment is visually clean (when observed under the same lighting and distance), I can be assured that the residue level on the surface is below X mcg/cm^2. I don't necessarily know the exact or precise value of that residue on a visually clean surface, but I know it is *below that VRL value*. Using this criterion, then the VRL is the lowest spiked level where I can see at least some amount of residue across the *entire* spiked surface.

To many this seems counterintuitive; shouldn't we be looking at the other end of the spiking levels, and either select the highest level at which the coupon was completely visually clean or the lowest level at which we can see only a small amount of residue on a portion of the spiked area. Let's look at the first situation (where the coupon was completely visually clean). Let's say that the spiked level was 0.5 mcg/cm^2; does this mean that any surface that is visually clean has a residue level below 0.5 mcg/cm^2? Well, the answer to this depends on whether we also spiked at a level of 0.6 mcg/cm^2. If we didn't spike at that level, then perhaps that higher level would also be visually clean; therefore to say that visually clean means it is below 0.5 mcg/cm^2 is a faulty conclusion. Okay, what if we spike at 0.6 mcg/cm^2 and it was in fact *not* visually clean; does this mean we can say any surface that is visually clean is below 0.5 mcg/cm^2? The answer is still "NO". Perhaps if we spiked at 0.55 mcg/cm^2 the surface would be visually clean. This logic can be continued to the point where we spike at a level very close to 0.5 mcg/cm^2, such as 0.51 mcg/cm^2.

There also is a problem with selecting the VRL as the lowest level at which there is at least *some visible residue on a small portion of the spiked area*. The issue is this: If we spike a coupon at a nominal level of 0.5 mcg/cm^2 (as an example), what does it mean if a portion of the coupon is visually clean and a portion is not visually clean? Well, it probably means that the residue is not evenly distributed over the coupon. It might be the situation where the visually clean portion is 0.45 mcg/cm^2 and the visually soiled portion is 0.55 mcg/cm^2. If that were the case, what conclusion can I draw about the spiked level of 0.5 mg/cm^2? How can it happen that we spike at one level and the coupon has different levels of residue on different portions of the surface? Very easily. Let's say I spill my coffee on a white table, and it forms a natural circle; as it dries, I will typically see a darker portion (higher levels of residue) around the outer portion of the circle. This probably happens because I get different rates of drying on the edges (as compared to the center portion), which could allow for some concentration gradients to form within the spiked solution (this sounds reasonable to me, but a physical chemist may have a better explanation of the phenomenon). So trying to draw a conclusion on spiked coupons which are only partially visually clean is not an acceptable practice.

This is why I advocate setting the VRL as the lowest level where I can see residue across the entire spiked surface (even if I can visually see different levels across the coupon). Remember that this VRL is a level where, if a surface is observed under the same conditions and is visually clean, the surface has a residue value below the VRL.

For clarification, the VRL limit for a given residue is not fixed. It depends on the viewing conditions (including lighting, distance, and angle of viewing). It also depends on the surface MOC; a white residue on PTFE will generally have a higher VRL as compared to the same white residue on stainless steel. Furthermore, as a practical matter, we may initially spike at levels of 0.25, 0.50, 1.0 2.0, and 4.0 mcg/cm², and determine that the VRL is 1.0 mcg/cm². If our calculated carryover residue limit were to be 0.70 mcg/cm², then a VRL of 1.0 mcg/cm² would not be adequate to confirm a visually clean surface was below the carryover limit (the best we could do was to say the visually clean surface was below 1.0 mcg/cm²). However, in that situation we have the option of doing *additional spiking levels* between 0.50 and 1.0 mcg/cm², which *might* result in a VRL of 0.70 or even 0.60 mcg/cm², resulting in a *useful* VRL; on the other hand, with those additional spiking levels, the VRL might be confirmed at a value of 0.80 mcg/cm² or above, in which case we could *not* claim that a visually clean surface has residue values below the calculated carryover limit of 0.70 mcg/cm². There clearly is an option to only do spiking levels down to a calculated carryover limit; others may choose to drive the VRL value as low as possible to account for possible future changes (that is, lower L3 values) in the calculated carryover limit.

This chapter does focus on why the "endpoint" for establishing VRLs should be the *lowest* spiked level at which a residue is visually seen across the *entire* spiked area. It sets the stage for Chapter 28, when we will take up the issue of applying VRLs to routine monitoring (that is, monitoring a cleaning process *after completion* of the validation protocol as part of "validation maintenance").

The above chapter is based on a *Cleaning Memo* originally published in November 2020.

REFERENCES

FDA, "Guide to Inspections Validation of Cleaning Processes", United States Printing Office, 1993.

Forsyth, R.J. "Visible-Residue Limit for Cleaning Validation and its Potential Application in Pharmaceutical Research Facility". *Pharm. Tech.* 2004, 28, p. 10.

Forsyth, R.J. "Application of Visible-residue Limit for Cleaning Validation in a Pharmaceutical Manufacturing Facility". *Pharm. Tech.* 2005, 29(10), pp. 152–161, October.

Fourman, G.L.; Mullen, M.V. "Determining Cleaning Validation Acceptance Limits for Pharmaceutical Manufacturing Operations". *Pharm. Tech.* 1993, 17, pp. 54–60.

LeBlanc, D.A., ""Visually Clean" as a Sole Acceptance Criteria for Cleaning Validation Protocols", *J. Pharm. Sci. Tech.* 2002, 56(1), pp. 31–36, January–February.

LeBlanc, D.A., "Understanding and Applying 'Visually Clean'", in *Cleaning Validation: Practical Compliance Solutions for Pharmaceutical Manufacturing, Volume 1*. Parenteral Drug Association, Bethesda, MD, 2006, pp. 207–209.

LeBlanc, D.A., "Visually Clean and Visual Limits", in *Cleaning Validation: Practical Compliance Solutions for Pharmaceutical Manufacturing, Volume 3*. Parenteral Drug Association, Bethesda, MD, 2013a, pp. 149–152.

LeBlanc, D.A., "More Uses for Visual Limit Determination", in *Cleaning Validation: Practical Compliance Solutions for Pharmaceutical Manufacturing, Volume 3.* Parenteral Drug Association, Bethesda, MD, 2013b, pp. 153–155.

LeBlanc, D.A., "Special Cases in Determining 'Visually Clean'", in *Cleaning Validation: Practical Compliance Solutions for Pharmaceutical Manufacturing, Volume 4.* Parenteral Drug Association, Bethesda, MD, 2017, pp. 133–135.

PIC/S, "Recommendations on Validation Master Plan, Installation and Operational Qualification, Non-Sterile Process Validation Cleaning Validation", Pharmaceutical Inspection Convention Pharmaceutical Inspection Co-Operation Scheme Document PI 006-1, Geneva, Switzerland, 2001.

28 Visual Residue Limits
Part 2

In this chapter, we discuss using Visual Residue Limits (VRLs) as part of *routine monitoring* to establish that a *quantitative* residue limit is being met. This is in line with the recent (2018) EMA Q&A that deals with using a visually clean criterion in routine monitoring to establish that a *quantitative* residue limit is being met [EMA, 2018]. If you haven't read (and understood) Chapters 26 and 27, please do so before considering this issue.

Before I start, here is a little history about the use of VRLs in validation (qualification) protocols. The PIC/S Cleaning Validation Recommendations [PIC/S, 2001] has a statement in its section on limits that the *most stringent* of three criteria should be used for limits *in a protocol*. Those three criteria are a dose-based calculation, 10 ppm in the next product, and visually clean. Since this PIC/S document was written before the advent of HBELs, it is reasonable that an HBEL criterion could be substituted for (or added to) the dose-based criterion. The idea of visually clean *alone* has not commonly been used because many critical locations in equipment cannot be readily viewed (such as transfer pipes), as well as for other reasons (such as the variability of viewing conditions and the variability of the viewers). We should also clarify that in situations where an *analytical method was required* to confirm that residues of the active were acceptable, it was still a requirement that the equipment be visually clean. The reason for that is that residues of *other* materials (such as excipients) could also cause the equipment to not be visually clean, and it is considered a reasonable GMP interpretation that equipment be cleaned to the point that there are no visible residues. Now let's move to the issue of using a visually clean criterion in *routine monitoring*.

Again, some more history. It has always been an expectation *following* a validated cleaning process that equipment be visually clean. That is *not* new. (Note that for "minor cleaning" within a campaign there is *not necessarily* an expectation that the equipment be visually clean. While that "minor cleaning" is *part* of an overall validation program, it *by itself* is not a validated cleaning process.) However, the purpose of establishing that the equipment was visually clean for routine manufacture was *not* to provide "clear" evidence that the residue limits were being met. The evidence that residue limits were being met was the original validation protocols combined with the fact that the cleaning procedures (SOPs) were carried out correctly by trained operators. Verifying that the equipment was visually clean on a routine basis was done more for the reason that if the equipment was not visually clean, then something was wrong. The fact that the equipment was not visually clean does not tell us about the nature of the residue or about the cause of the residue. An investigation is needed as part of a CAPA program.

One way to think of the traditional use of this requirement (that equipment be visually clean following routine cleaning) is that in such situations "visually clean" is a *necessary requirement* to say that the cleaning process was carried out correctly, but it (alone) is *not a sufficient reason* to say the cleaning process was carried out correctly. If the equipment is in fact visually soiled after the validated cleaning process, that (alone) is *sufficient* reason to say something is wrong, but it doesn't tell us what is wrong (or provide the cause of the problem).

Now we get to the issue in Question #7 of the 2018 EMA Q&A document. That question and answer are as follows:

> Q7. Is analytical testing required at product changeover, on equipment in shared facilities, following completion of cleaning validation?
> A: Analytical testing is expected at each changeover unless justified otherwise via a robust, documented Quality Risk Management (QRM) process. The QRM process should consider, at a minimum, each of the following:
> - the repeatability of the cleaning process (*manual cleaning is generally less repeatable than automated cleaning*);
> - the hazard posed by the product;
> - whether visual inspection can be relied upon to determine the cleanliness of the equipment at the residue limit justified by the HBEL.

Note that this clearly is in the context of *analytical* testing for *routine monitoring* ("following completion of cleaning validation"). The first bullet point suggests that this may be more critical for *manual cleaning* processes. The second bullet point suggests that it may be more critical for *low PDE actives*. The third bullet point suggests that the context might be residues (such as the active) for which an HBEL has been determined *for the validation protocol*. It further suggests that analytical testing is expected *unless* there is a risk assessment to justify *not* doing so. Different companies may have different approaches to address this issue for routine monitoring. Here are some possibilities:

1. For low PDE products, perform at least one chemical analytical test for the active as part of routine monitoring. This does not have to be with the same level of sampling as in the validation protocol. For example, for automated cleaning processes where a final rinse sample is a *reliable indicator* of possible residue levels, just a single final rinse sample for the low PDE active may be adequate. Of course, if this is an automated cleaning process, a final rinse conductivity is also still appropriate as an indicator of *process control*; and a visually clean requirement is also still appropriate. If this is *manual* cleaning for a low PDE active, a swab sample on one or more *critical* (worst-case) sampling locations may be adequate (still with the requirement that the equipment be visually clean).

2. For low PDE products, if the preference is not to perform a chemical analytical test for the low PDE active, then this is when it is *necessary* to do spiking studies (*as discussed in Chapter 27*) to determine what the VRL is.

If the VRL is the same or above the calculated carryover limit for that active, then a visual examination alone may establish that the residue is below the calculated limit. Note that depending on the active, it is entirely possible that this desirable relationship between the VRL and the calculated limit would not hold; in which case, using option #1 above should be considered.

3. For products with actives that do *not* have a low PDE (which I will call "traditional" actives), the same two options above can be considered. However, in the case of "traditional" actives, it is more likely that the VRL will be above the calculated limit (spiking studies could be used to determine the VRL). However, some companies may choose to come to this conclusion *without* performing spiking studies. For example, it is highly likely that a *calculated* L3 residue limit (amount per surface area) of 4 mcg/cm² would be readily visible on equipment surfaces. Some companies might want to use a lower value for comparison to the calculated limit, such as 1 mcg/cm², while others might want to use a higher value of 10 mcg/cm² for comparing to the calculated limit (remembering that the lower the *calculated* limit, the less likely that "visually clean" will be suitable confirmation of meeting that residue limit).

An additional concern related to visual examination for *routine monitoring* is whether the equipment *must* be disassembled *to the same ext*ent as disassembly during the cleaning validation protocols (so that the identical surfaces can be observed). This may be possible; however, I would carefully consider whether product quality is being impacted more by the possibility of recontamination by the disassembly/visual assessment/reassembly process as compared to the situation where the visual examination is limited to what might be "practical". Each company will have to consider this as part of a risk assessment as to the extent of disassembly. For this consideration of the extent of disassembly, the EMA's answer to Question #8 in the 2018 EMA Q&A document should also be consulted. Included in that answer is the following dealing with disassembly:

> Visual inspection should include all product contact surfaces where contamination may be held, including those that require dismantling of equipment to gain access for inspection and/or by use of tools (for example mirror, light source, boroscope [sic]) to access areas not otherwise visible.

In exploring these options for your specific situation, please be aware that this discussion is in the context of VRLs as defined in Chapter 27. If you have a *different* definition of VRL (or a different means of establishing VRLs), then this discussion might not apply. In any case, these considerations may be addressed in appropriate risk assessments as part of a "life cycle" cleaning validation approach for routine monitoring.

The above chapter is based on a *Cleaning Memo* originally published in December 2020.

REFERENCES

European Medicines Agency, "Questions and Answers on Implementation of Risk Based Prevention of Cross Contamination in Production and 'Guideline on Setting Health Based Exposure Limits for Use in Risk Identification in the Manufacture of Different Medicinal Products in Shared Facilities' (EMA/CHMP/CVMP/SWP/169430/2012)". Document EMA/CHMP/CVMP/246844/2018. 19 April 2018.

PIC/S, "Recommendations on Validation Master Plan, Installation and Operational Qualification, Non-Sterile Process Validation Cleaning Validation", Pharmaceutical Inspection Convention Pharmaceutical Inspection Co-Operation Scheme Document PI 006-1, Geneva, Switzerland, 2001.

Section V

Analytical and Sampling Methods

The following four chapters deal with issues related to analytical methods, sampling methods, and sampling recovery studies.

DOI: 10.1201/9781003366003-33

29 Two More Nails in the Coffin?

I have for years taught (or preached) that performing *quantitative* swab recovery studies at multiple spiked levels is a waste of effort and resources [LeBlanc, 2010; LeBlanc, 2013a; LeBlanc, 2013b; LeBlanc, 2017]. Furthermore, I have also maintained that, other things being equal, the percentage recovery of spiked material will decrease as the spiked level increases. If this is true, then one spiked level at the acceptance limit should be the *same or a smaller percentage recovery* as compared to lower spiked levels, which is where you are typically expecting your residue values to be in a protocol.

Two recent papers also have data that seem to support what I teach (although one explicitly questions my approach). Hence, the title of the chapter, "Two More Nails in the Coffin"; that said, I realistically expect the idea of *multiple* spiked levels for recovery studies will still be around for many years in the future.

One study is published online [Gerostathi et al., 2019]. This is the study that questions my approach, where my approach is called a "misconception". While the data generated in that publication does not appear to be developed to disprove my contention, it actually *supports* my contention. Here is the data presented for Case I (which can be viewed online).

Quantity Spiked on Coupon (ppm)	Quantity Recovered from Coupon (ppm)	Percent Recovery (%)
50	58	116.6
75	68	90.7
100	80	80.0
125	94	75.2
150	105	70.0

Now, this data would seemingly support my contention that percent recovery decreases with increasing spiked levels. However, the equation associated with this data to give a linear response was given by the authors as $Y = 33.53 + 0.4752X$ (with Y being the "Percent Recovery" and X being the "Quantity Spiked"). What this equation means is that if the spiked amount was "zero", the percent recovery was about 33% (which should *not* be the case); the high value for what could be the "blank" suggests that there is something amiss. However, in any case, the authors plotted Quantity Recovered (Y-axis) vs. Quantity Spiked (X-axis) and achieve a straight line (makes sure you don't confuse "X/Y-axis" with the "values of X and Y" in the equation). An *appropriate* interpretation of this data is more

DOI: 10.1201/9781003366003-34

correctly that the *slope* of the straight line in Figure 1 in the publication is the recovery percentage; therefore over that narrow spiked range of 50–150 ppm, the recovery percentages are essentially the same (about 43%). In other words, the data in Case I in this online publication is amazingly consistent with what I have taught (that over a narrow range, recovery percentages will be the same within experimental variation).

The second study was published in *PharmTech* magazine and is titled "Qualification of a Swab-Sampling Procedure for Cleaning Validation" [Ovais et al., 2020]. This publication presents data looking at the different variables in swab recovery studies, including direction of swabbing strokes, swabbed area, swab type, solvent concentration for wetting swabs, surface finish, spiked amount, and personnel doing the swabbing. Note that the spiking material was a detergent (Teepol), so my expectation would be that recoveries should be relatively high. I had some problems understanding what the authors actually did (the amounts spiked were reported in Table II in the publication as 90 or 110 micrograms for either 20.25 or 30.25 cm^2, while the text said the spiked levels were 50, 100, and 200 mcg/cm^2). However, the key item for this chapter is that the authors report that for Sampler I, "the average recoveries could also be seen decreasing with increasing spiked amounts", which might be seen as consistent with my view. However, I don't believe that the data is supportive of my contentions about the relationship between spiked level and percent recovery. While there were differences between the two samplers, the overall mean pooled recovery percentages were in the 85–115% values as expected for a spiked material that should have a high recovery. As I look at the results of this study, what it shows is the personnel doing the sampling are the most important source of variability. Yes, we should try to control the variables as much as possible; but if we are concerned about the variability of swab sampling, we should design our cleaning process to produce residue levels *significantly* below our acceptance limits.

Let me make it clear that I have no *well-designed* studies that show within *narrow* ranges the percentage recovery is essentially the same, or that over *wide* ranges the percentage recovery decreases with increasing spiked levels. I should clarify this last approach by stating that the two *main* variables (other than the person doing the swabbing) that contribute to the variability of results in recovery studies are the variability of the swabbing itself (mainly the person doing the swabbing) and the variability of the analytical method. There may be other issues that contribute to variability, such as the variability of the preparation of the spiked coupon. It should be clear that generally the lower the spiked level, the more potential variability there is due to the analytical method itself. What this means is that at very low spiked levels, I *may* find that the measured percentage recovery is lower than at higher levels. This may particularly be a problem when recovery studies are done near the LOQ of the analytical method. Furthermore, if the spiked amount is at the LOQ, it becomes almost impossible to determine a *reliable* quantitative percentage recovery unless the recovery is actually 100%.

With all these caveats, I am still open to being disproved. I am more than willing to work with anyone to design a protocol that might disprove my contentions. If you are interested, you should have my contact information.

The above chapter is based on a *Cleaning Memo* originally published in October 2020.

REFERENCES

Gerostathi, I.; Ovais, M.; Walsh, A. "Recovery Studies: Common Issues & Using Statistical Tools to Understand the Data". 18 September 2019. https://www.pharmaceuticalonline.com/doc/recovery-studies-common-issues-using-statistical-tools-to-understand-the-data-0001 (accessed May 6, 2021).

LeBlanc, D.A., "Spiking Amounts for Sampling Recovery Studies", in *Cleaning Validation: Practical Compliance Solutions for Pharmaceutical Manufacturing, Volume 2*. Parenteral Drug Association, Bethesda, MD, 2010, pp. 137–139.

LeBlanc, D.A., "Swab Sampling Recovery as a Function of Residue Level", in *Cleaning Validation: Practical Compliance Solutions for Pharmaceutical Manufacturing, Volume 3*. Parenteral Drug Association, Bethesda, MD, 2013a, pp. 181–185.

LeBlanc, D.A., "Revisiting Linearity of Swab Recovery Studies", in *Cleaning Validation: Practical Compliance Solutions for Pharmaceutical Manufacturing, Volume 3*. Parenteral Drug Association, Bethesda, MD, 2013b, pp. 177–179.

LeBlanc, D.A., "Revisiting Linearity of Recovery Studies", in *Cleaning Validation: Practical Compliance Solutions for Pharmaceutical Manufacturing, Volume 4*. Parenteral Drug Association, Bethesda, MD, 2017, pp. 143–145.

Ovais, M.; Huey, Q.C.; Ngadinan, M.; Shamungam, T. "Qualification of a Swab-Sampling Procedure for Cleaning Validation". *Pharm. Tech.*, July 2020. https://cdn.sanity.io/files/0vv8moc6/pharmtech/36af3ab6a25e429675c3cbbe93f132983e43d479.pdf (accessed May 6, 2021).

30 Timing for Swab Sampling in a Protocol?

In performing swabbing *for chemical residues* after completion of a cleaning process in a cleaning validation protocol, is the timing for taking the swab sample critical? In the tradition of being a consultant paid by the hour, the *correct* answer is "It depends".

When I do swab sampling at the end of a cleaning process, I will usually wait until the equipment is dry. One reason for waiting until the equipment is dry is that I may also do a visual examination at the same time; a visual examination under dry conditions is usually more stringent than a visual examination of wet equipment. Another reason for waiting until equipment is dry before starting swabbing is that recovery studies are usually done on dry coupons, so I would want my swab sampling in a protocol to reflect that same condition.

Another issue that may cause a delay in swab sampling of the surface is the traditional conflict between people who are doing swabbing for *microbiological* purposes and those who are doing swabbing for *chemical* residues. This conflict revolves around which is done first. The argument for doing microbiological sampling first is that it is that the people swabbing for chemical residues may not be aware of or sensitive to microbiological contamination issues. The argument for doing swabbing for chemical residues first usually involves the use of TOC as the analytical method. If after microbiological sampling the surface is "cleaned" with an alcohol/water blend, the air in the environment around the sampling location may have levels of the volatile alcohol which could cause high TOC values in the swab sampling. The purpose of this chapter is not to resolve this "conflict", but to say that proper training and proper procedures should help resolve those concerns.

In general, the issue of the timing of swab sampling for chemical residues should be that sampling should be as soon as practical. For example, it is always appropriate from an operator safety perspective for equipment rinsed at elevated temperatures to wait until the equipment cools down considerably. In addition, swabbing as soon as practical *avoids* the situation where, despite all the warning placards about the validation status, the equipment is used for product manufacture *before* swab sampling is performed.

This brings us to the main emphasis of this chapter. It is possible that residues on surfaces *may* degrade considerably due to exposure to such elements as light or air. If such degradation *after* completion of the cleaning process is real, we are now talking about a real concern. That concern is more valid if the analytical method used is a specific one, such as HPLC. The longer we wait to swab sample the surface, the lower our analytical values might be. The end result may be that

DOI: 10.1201/9781003366003-35

we pass the acceptance criterion when it would have failed if we had swabbed *earlier*. Now it might be countered that if we sample immediately after the cleaning process, the active will still degrade on the equipment before the next use. And then we will have to evaluate the safety of that degradant.

For the vast majority of cases, it is probably the case that this degradation issue does not apply. So, let's return to the main issue. And the answer is that we probably *already* have data showing the maximum acceptable delay time for swabbing for chemical residues. When a swab recovery study is done, there is usually an interval between the time of spiking the chemical material and the time of swabbing the coupon. That time could be the starting point for setting a maximum "hold time" for sampling. Some may record the end of the cleaning process as the time at the end of draining of the equipment, some might record it as the end of an airblow, and some might record it as the end of drying. Obviously, if in a sampling recovery study the chemical material is spiked with a volatile organic solvent, the time from spiking until dryness is minor as compared to spiking with an aqueous solution. In any case, the time between dryness on the coupon and initiation of swabbing in a recovery study may provide a reasonable estimate of the maximum time that should be allowed for swab sampling in a cleaning validation protocol for the cleaning process itself. That is, if I allow coupons to dry for 18 hours in a recovery study, then in a cleaning validation protocol, I should perform swab sampling within 18 hours.

There may be other considerations, such as whether the equipment is maintained under nitrogen after cleaning, thus reducing the likelihood of any degradation. Furthermore, if I have concerns about degradation on surfaces after cleaning, I might use TOC as my analytical method (assuming the chemical species does not degrade all the way to carbon dioxide). Finally, I might decide to do formal studies and evaluate residues on a spiked surface as a function of time. In such a study, I could either prepare replicates at an L3 (amount per surface area) level or swab *different* replicates over a certain time (such as over four days). Such a formal study may be useful to provide additional support for my maximum "hold time" before swabbing.

Note that this issue of degradation after completion of the cleaning process does not apply to rinse sampling *if* the sample is the last portion of the final process rinse. However, if the rinse sample is a *separate* sampling rinse after completion of the process rinse, then similar principles are generally applicable.

The above chapter is based on a *Cleaning Memo* originally published in October 2018.

31 More Swab Sampling Issues

Chapter 30 addressed the issue of "timing" in performing swab sampling in a protocol. This chapter considers four miscellaneous topics in swab sampling.

CONTROLLING THE SWABBED AREA IN PROTOCOLS

There are basically two techniques used to control the swabbed area *in protocols*. One is what I call the "eyeball" method. That is, the operators are trained to swab a defined area (plus or minus) *without* any template. Companies may utilize a diagram (typically 25 or 100 square centimeters) in the swab SOP to assist in that process. Another option to assist in the eyeball method is to have a coupon (preferably a translucent plastic) cut to the desired surface area. These adjuncts are *never* placed on the surface but may be held near the surface to assist in swabbing the correct area.

A second option is to use a template *placed on the surface* to be sampled. Traditionally, a template is a plastic sheet with a defined area (typically 25 or 100 square centimeters) cut out *within* the template. The template is placed on the surface to be sampled, and the swabbed area is controlled by the area of the cut-out section of the template. A possible variation in the template method is to have a template with an area *slightly larger* than the area to be swabbed. For example, if my swabbed area was to be 10 cm × 10 cm, the template might have a cut-out area of 11 cm × 11 cm. That way, once I place my template on a surface, my swab strokes would stop about 0.5 cm from the edge of the template. There is still some element of variability as in the "eyeball" method, but the variability is significantly reduced. This *larger* template avoids the issue of the swab head contacting the template edge, thus avoiding leaving behind some liquid at the surface/template interface.

My recommendation between the eyeball and template method has generally been to use the eyeball method. The reason is that the worst-case swabbing locations are typically *not* flat surfaces. Templates can be problematic if the surface is not flat (although they may work on gently curved surfaces if the template is sufficiently flexible). That said, if all sampled surfaces are flat, the template certainly controls the swabbed area better, and the use of the slightly larger template area avoids contact of the swab head with the template itself.

DOI: 10.1201/9781003366003-36

CONTROLLING THE SWABBED AREA IN RECOVERY STUDIES

The same two methods discussed above for protocols can also be used *for recovery studies* for *larger* coupons. However, there are two more possibilities to consider for recovery studies. One is to use a coupon the exact size of the area to be swabbed. That is, if the swabbed area is 100 square centimeters, my coupon is cut exactly 10 cm × 10 cm. The main concern with this option is having the swab head touch the coupon edges, which can lead to loss of liquid and a corresponding low recovery percentage. This issue can be overcome by having a coupon slightly larger than the area to be swabbed (much like having a slightly larger area for the template discussed above). For example, I might use a coupon that is 11 cm × 11 cm. When I swab the coupon, I would stop my swab strokes about 0.5 cm from the edges of the coupon. That way I can control my swabbed area reasonably well without having the swab head touch the edges of the coupon.

USE OF AN EXTENSION POLE FOR SWABBING INACCESSIBLE LOCATIONS

This is one of my *major* "pet peeves". I know many companies use this technique, but I see little reason for its use in most cases. If I have a swab on the end of a 3-meter pole, how do I control the pressure applied to the swab head? And, how do I control the surface area sampled? Furthermore, if I am sampling through an opening in the top of a large vessel, I may be able to sample the sidewalls of the vessel, but how do I sample *under* the agitator blades, certainly a more significant worst-case location. In such a situation, unless I am willing to have a person enter the tank for swab sampling, I would much prefer to depend on rinse sampling (by setting limits appropriately for my rinse sample and by having recovery studies done for the rinse sampling procedure).

USE OF MULTIPLE SWABS FOR THE SAME SURFACE

The issue here is that I may be able to increase the recovery percentage by swabbing the *same* surface twice (or even thrice). That is, I use one swab first over a specified area (for example, a 10 cm × 10 cm surface) with that first swab being a wetted swab. Then I swab that same 10 cm × 10 cm surface with a second swab, and place both swabs in the same vial for extraction. In some cases, the second swab may be a dry swab, based on the fact that the second swab will "mop up" the liquid left on the surface from the first swab. In other cases, I might want that second swab to also be a wetted swab, based on the possibility that the limiting factor is the dissolution of the residue onto the swab (with a second wetted swab allowing a greater percentage recovery).

If I wanted to investigate whether two swabs give me a significantly better recovery, there are two possible options. One is to perform a recovery study with two swabs, but by putting the swabs in separate vials for separate extraction and analysis. The second option would be to do two separate coupons (spiked the same), one coupon with one swab and a second coupon with two swabs, and then compare the recoveries.

My preference for a given facility is to *always* use one swab for sampling, or to *always* use two swabs. While it is possible to set up a sampling SOP using one swab for certain specified products and two swabs for other specified products, this requires more attention and care to make sure the correct number of swabs is used in a given protocol.

If multiple swabs are used for a given surface area, then it is important, particularly if TOC is the analytical method, to make sure that my blank is also with multiple swabs.

This chapter does not exhaust all the concerns with swab sampling, but it does point out concerns and options with certain practices.

The above chapter is based on a *Cleaning Memo* originally published in November 2018.

32 Pass/Fail Analytical Test Methods

A pass/fail test method is an analytical test method that only tells me whether I am below a certain value. For example, I may have an HPLC method that strictly tells me that I am *at or above* a value of X ppm in my sample, or I am *below* a value of X ppm. In the context of cleaning validation, if the limit in my analytical sample (a swab extract, for example) is X ppm, that pass/fail test method should allow me to say whether I am meeting my acceptance criterion, assuming that my sampling recovery is 100% or that I have dealt with a lower recovery by other means (as discussed below).

This pass/fail method may be a *detection limit* test procedure. That is, I am strictly measuring whether I am below the detection limit. If the detection limit is at or above the acceptance limit in the protocol, demonstrating that I am *below* that detection limit should be evidence that I am meeting my protocol acceptance criteria. Another type of pass/fail test method is one where I am comparing the measured response *of the test sample* to the measured response *of a standard* representing the 100% concentration (ppm) of my acceptance limit. In other words, if the acceptance limit is X ppm of my active, I prepare a standard at X ppm of that active. I run that standard and my test sample in my HPLC procedure, and compare the responses (area under the curve or peak height). If the response of my test sample is less than the response of my standard, then I have demonstrated that the test sample is meeting my acceptance criterion. Note that, generally, that standard is carried along for *every set* of test samples that are analyzed. I may also prefer to run a standard *before* and *after* my test samples (that is, I bracket my test samples with injections of my standard).

This is generally *not* what most companies do for analytical methods for cleaning validation purposes. Most companies prefer to have a quantitative method that is validated to be accurate and precise *over a certain range*. One reason for doing so is to demonstrate the robustness of the cleaning method. For example, if my limit is X ppm, I have much more confidence in the robustness of my cleaning process if I have data that shows samples are consistently in the range of 0.1X–0.3X ppm as compared to just being able to say that I am only below X ppm (as in a pass/fail method based on the LOD). A second reason for having a method validated over a certain range is if I am using a "stratified sampling" approach to setting limits [LeBlanc, 2013a; LeBlanc, 2013b; LeBlanc, 2013c]. However, the point is that at least from a scientific point of view, a pass/fail method may be acceptable. One situation where a pass/fail method might be used is in a cleaning verification mode for clinical trial material manufacturing, where spending time and effort for a method validated over a range may not be required.

DOI: 10.1201/9781003366003-37

Now we'll cover one of the main concerns in the use of a pass/fail test method, the issue of dealing with *sampling recovery*. That is, my pass/fail point is X ppm; but suppose my recovery might be as low as 50%. If my swab test sample shows that I am below X ppm, how can I say I am meeting my acceptance limit if my sampling recovery is only 50%, or for that matter any recovery percentage value below 100%?

What can I do to avoid this recovery problem with the pass/fail test method? The simplest way is to just set the pass/fail criterion in the test method at something below the calculated acceptance limit in the analytical sample. For example, if my calculated swab limit is X ppm, I can assume a worst-case sampling recovery of 50% and establish my pass/fail value in my test procedure as 0.5X ppm. Note that this doesn't work if my pass/fail value is solely a true detection limit (LOD) value; in that LOD situation, I cannot establish that I am meeting my acceptance criterion if my sampling recovery is less than 100%.

But, you ask, what if my sampling recovery is actually less than 50%? Don't I have a problem there? The answer is that I don't just *assume* a 50% (or higher) recovery. I do something to *demonstrate* that my recovery is *at least* 50%. How is that done? Very simply. I first establish my pass/fail method based on a value of 0.5X ppm rather than X ppm. Then I perform a recovery study by spiking at a level representative of the value of the X ppm limit. I then perform my analysis on that test sample with my pass/fail test procedure. As long as my result shows a value at or *above* 0.5X, then I have clearly demonstrated that my sampling recovery is *at least* 50%. In this example, if I spiked at an equivalent of X ppm, and the measured value in my recovery study was *below* 0.5X ppm, I would have demonstrated that my sampling recovery in this case was *less than* 50%. Therefore, in the latter case, using that combination of a pass/fail value of 0.5X ppm and the sampling method would be *inadequate* to demonstrate that I was meeting my acceptance criterion of X ppm.

In the example presented above, the pass/fail was established at 0.5X ppm. If I had a requirement that I must have a recovery of at least 50%, I could establish a pass/fail value of 0.5X ppm, or I could establish a higher pass/fail value. That is, I could still spike my coupon at X ppm, and then analyze with my pass/fail method at 0.7X ppm. If I showed a value of 0.7X ppm or greater, then I have established a recovery of *at least* 70%. Therefore, the pass/fail method at 0.7X ppm could be used.

A second issue in the use of a pass/fail test method relates to demonstrating the robustness of my cleaning procedure. That robustness may be established by setting the pass/fail point at a value *lower* than the calculated limit. For example, that lower value may be 0.8X ppm or 0.6X ppm. Note that this would be *in addition to* dealing with the sampling recovery issue by using a lower value.

Note further that this establishment of an adequate recovery will generally only work in cases where the pass/fail limit is established by running a *known standard* and that an adequate measured response is obtained at a level below the ppm acceptance limit. In cases where the pass/fail point is a *true* detection limit, this method of establishing adequate recovery will not work.

One last discussion point is that a pass/fail method still requires analytical method validation. However, the method validation for a pass/fail method is generally less time-consuming as compared to a method validated over a wide range.

The point of this chapter is not to recommend the use of a pass/fail analytical procedure. Rather it is to present conditions to reliably and appropriately use such pass/fail methods in validation and/or verification protocols.

The above chapter is based on a *Cleaning Memo* originally published in June 2017.

REFERENCES

LeBlanc, D.A., "Basics of 'Stratified Sampling'", in *Cleaning Validation: Practical Compliance Solutions for Pharmaceutical Manufacturing, Volume 3*. Parenteral Drug Association, Bethesda, MD, 2013a, pp. 93–96.

LeBlanc, D.A., "More on 'Stratified Sampling'", in *Cleaning Validation: Practical Compliance Solutions for Pharmaceutical Manufacturing, Volume 3*. Parenteral Drug Association, Bethesda, MD, 2013b, pp. 97–100.

LeBlanc, D.A., "Final Notes on 'Stratified Sampling'", in *Cleaning Validation: Practical Compliance Solutions for Pharmaceutical Manufacturing, Volume 3*. Parenteral Drug Association, Bethesda, MD, 2013c, pp. 101–104.

Section VI

Product Grouping

The following three chapters deal with issues related to product grouping for protocols.

DOI: 10.1201/9781003366003-38

33 Issues in Product Grouping

This is part of a clarification on a topic I have written about in the past in a previous volume in this series [LeBlanc, 2006], but where I have subsequently refined and elaborated in many training seminars and webinars. So this is an opportunity to share that additional information with those that have not attended my seminars or webinars.

Product grouping (also called matrixing, family approach, or bracketing) involves selecting a representative product among a group, performing a cleaning validation protocol on that representative product, and applying the validation result to all products in the group. Products in a group are usually the same type of product (liquid, solid, or semisolid) manufactured on the same or equivalent equipment, and cleaned by the same cleaning process. Ordinarily, the representative product is the worst-case product. I generally identify the worst-case product as the product that is "most difficult to clean". That is, if I can clean the "most difficult to clean" product acceptably, the other products in the group should also be cleaned acceptably.

Part of the criteria for determining that the worst-case product is cleaned acceptably is *setting residue limits for that worst-case product*. That is, let's assume I have products A, B, C, and D in a group. I determine (by whatever criteria) that A is the worst case (most difficult to clean). What limits do I establish in my protocol when I perform my protocol on A? Do I calculate my residue limit for A based on carryover calculations utilizing each of the other products (B, C, and D) as the next product, and then select the lowest of those possible limits as the limit for my protocol? The issue with such an approach is that it may be appropriate in some situations, but *not* in other situations. Here are several examples to illustrate the issue.

Example 33.1

Suppose that A is the worst-case product (most difficult to clean). Here are my calculated limits in μg/cm² for the cleaning for each of A, B, C, and D (based on the possibility of each of the other products being the *next* product). In each case, the lowest limit for each cleaned product is *italicized*.

DOI: 10.1201/9781003366003-39

Cleaned Product	Limit (µg/cm²) with Specified Next Product			
	A	B	C	D
A	N.A.	1.5	*1.3*	1.6
B	*1.5*	N.A.	1.7	2.1
C	2.3	2.7	N.A.	*2.0*
D	*1.5*	1.9	2.3	N.A.

In this situation, if I cleaned A down to a limit of 1.3, then the other three products (which are easier to clean) should also be cleaned down to a limit of 1.3, which is *below the lowest limit* for each of those other three products. In this example, cleaning of A down to its limit of 1.3 provides assurance that the other three products are effectively cleaned at their calculated limits.

Example 33.2

In this example, we will still suppose that A is the worst-case product (most difficult to clean), but the calculated residue limits are *different* as given in the following table.

Cleaned Product	Limit (µg/cm²) with Specified Next Product			
	A	B	C	D
A	N.A.	1.5	*1.3*	1.6
B	*1.5*	N.A.	1.7	2.1
C	2.3	2.7	N.A.	*2.0*
D	*0.9*	1.1	1.4	N.A.

In this situation, if I cleaned A down to a limit of 1.3, then products B and C (which are easier to clean) should also be cleaned down to a limit of 1.3, which is below the lowest limit for each of those two products. However, applying that data to Product D is a problem. Product D is easier to clean than A, so it should be cleaned down to a limit of 1.3; however, I need to have Product D cleaned down to a level of 0.9, significantly *lower* than 1.3. Therefore, in this situation cleaning A down to its calculated limit (based on what is the worst-case next product) does *not* provide assurance that I have covered the situation of cleaning D down to its calculated limit.

In the early day of cleaning validation, the attempt to resolve this dilemma in a grouping approach was to perform *one protocol* on the worst-case (most difficult to clean) product and a *separate protocol* on the "most toxic" product. What was meant by "most toxic" product is the product with the lowest limit. So in the case of Example 33.2, I would perform one protocol on A at its limit of 1.3 and one protocol on D at its limit of 0.9. With these two protocols for Example 33.2, I have effectively shown that I can cover the cleaning validation of all four products by this strategy. However, while this strategy works well in the example given, let's take a look at another example where that approach is problematic.

Example 33.3

In this example, we will still suppose that A is the worst-case product (most difficult to clean), but the calculated limits are different as given in the following table.

	Limit (µg/cm²) with Specified Next Product			
Cleaned Product	A	B	C	D
A	N.A.	1.5	1.3	1.6
B	1.1	N.A.	1.4	1.5
C	2.3	2.7	N.A.	2.0
D	0.4	0.7	0.9	N.A.

For this example, let's suppose the *order of difficulty of cleaning* is as follows:

Most difficult: A
Second most difficult: B
Third most difficult: C
Fourth most difficult: D

For clarification, this means that D is the *easiest* product to clean.

Let's say in this example, I tried the approach of one protocol for the "most difficult to clean" product and one protocol for the "most toxic" product. Would this cover all four products? In this example, if I cleaned A down to a limit of 1.3, then product B (easier to clean) should also be cleaned down to a limit of 1.3; however, I need to have it cleaned down to a lower limit of 1.1. Can I then look to have B covered by a separate protocol for the "most toxic" product, which is D? If I can clean D to a limit of 0.4, what does it mean for B? The answer is that I *can't* come to a logical conclusion because D is *easier* to clean than B. In other words, just because D can be cleaned to a limit of 0.4, I can't say anything about what that means for B, even though the limit for B is *higher*. [For those of you who are wondering about C, it is easier to clean than A and has a higher limit; therefore cleaning of A at its limit is adequate to establish that C is effectively cleaned.]

ONE APPROACH FOR ALL EXAMPLES

The way to cover all of these examples given with *one protocol* is to perform the cleaning process on the "most difficult to clean" product using the *lowest limit of any product in the group*. In Example 33.1, this would mean a protocol with Product A at a limit of 1.3. In Example 33.2, this would mean a protocol with Product A at a limit of 0.9. In Example 33.3, this would mean a protocol with Product A at a limit of 0.4. The beauty of this approach is having one approach that covers all situations in one protocol.

Some may object that by including the toxicity or limit factor in the selection of the "worst-case" product, they have adequately addressed the issue of lower limits. That is, they may have used a "point system" which includes a "toxicity" factor (such as by a PDE value or by a limit calculation) as well as factors such as solubility of the active. While that may sound plausible, it still doesn't adequately and logically cover all situations. It is not unlike the old grouping approach of "most difficult to clean" and "most toxic", where it is possible to have some products "fall through the cracks". I should make it clear that my preference is *not* to use a toxicity factor in determining which product is "most difficult to clean". Toxicity may be a *risk* factor, but toxicity *per se* does *not* make something more difficult to clean (as discussed later in Chapter 34). As a risk factor, toxicity is addressed by selecting the lowest limit of any product in the group. The only exception to this preference is if I have two products which are *equally* difficult to clean, and one is highly hazardous and one is not highly hazardous. In that situation, I would prefer to use the highly hazardous product as the worst-case product in a grouping approach (the limit will be based on either the limit for that highly hazardous product or for another highly hazardous product).

Others may object that it may not be possible to analyze the "most difficult to clean" product at the lowest limit of any product in the group. That is a valid concern. If a better analytical technique cannot be found to overcome that objection, it may be necessary to divide the products into two (or more) groups to cover all situations. The approach in this situation (in terms of forming the two groups) will depend on the number of products, the relationship between the difficulty of cleaning among products, the calculated limits, and the available analytical methods.

The objective of this chapter is to present valid concerns for selecting products and limits for use in product grouping approaches for cleaning validation.

The above chapter is based on a *Cleaning Memo* originally published in May 2017.

REFERENCES

LeBlanc, D.A., "Product Grouping Strategies", in *Cleaning Validation: Practical Compliance Solutions for Pharmaceutical Manufacturing, Volume 1*. Parenteral Drug Association, Bethesda, MD, 2006, pp. 167–172.

34 Toxicity as a Worst-Case Grouping Factor

In a grouping (matrixing) approach to cleaning validation, one product is selected as the "worst case", and cleaning validation is done on that product. Successful validation of that product constitutes coverage of validation for all products in the group. My recommended approach is to select the "most difficult to clean" product in the group as the product to perform my validation protocols on, but to set limits for the active in that product at the lowest limit of *any* active for products in the group.

But, I often see an approach where a variety of factors are involved in selecting the worst-case product, and that among these factors is one item called "toxicity" (or a similar designation). In this approach, the toxicity may be based on PDE/ADE values, on a 1/1,000 dose criterion, or on other toxicity data. While this is a common practice, it is not one that I typically recommend. Why?

Perhaps, it is because I generally like to select the product I perform my protocol on as the "most difficult to clean" product. And I am not sure how the toxicity of a product makes it more (or less) difficult to clean (other than it has to be cleaned to a lower limit). Others may select the product to perform protocols on only as a "worst case" product, and using that "worst case" criterion (rather than my "most difficult to clean" criterion), I can understand why some may include toxicity as one of the factors in making a product a "worst case".

But is toxicity a relevant factor to consider in a grouping (matrixing) approach? Of course! This is why in my recommended approach I specify performing a protocol on the most difficult to clean product but setting limits for that product based on the *lowest limit* of any product in the group.

Here is an example that may illustrate a possible pitfall in a worst-case selection that includes toxicity as one of the determining factors for selecting the product to use for the protocol runs. Let's assume that we have two products (A and B) with the following characteristics, and let's assume we have only two factors in determining the worst case. The factors are *solubility* of the active and *PDE* value of the active. Let's assume that equal weight is given to each factor and that each is assessed on a scale of one to five (with higher numbers being the worst case.

Here is my example rating system.

Solubility (ppm)	Points	Toxicity (PDE) (µg)	Points
1–10	5	<1	5
>10–100	4	1–<10	4
>100–1,000	3	10–<100	3
>1,000–10,000	2	100–<1,000	2
>10,000	1	1,000–<10,000	1

DOI: 10.1201/9781003366003-40

Let's assume the "point" values for Product A and Product B are as follows:

	Solubility Points	Toxicity Points
Product A	5	1
Product B	2	5

That is, Product A is the least soluble and Product B is the most toxic.

In an approach where I either multiply the points or add the points, Product B is the worst case (most points). In this situation, I would then perform my validation protocol on Product B and set my limit based on the Product B limits. Note in this example that this limit would be used whether I selected either the limit of the worst-case product or the lowest limit among all products in the group. Further note that I am assuming the L3 limit (the limit per surface area) of the active in Product B would be lower than the L3 limit of the active in Product A.

The concern about this approach is "how do I know Product A would effectively be cleaned to *its* limit if, based on the solubility of the active, Product A was the least soluble (or 'most difficult to clean')?" The answer is "I don't know for sure, even though it *might* be the case".

The case presented is meant to be an extreme case for illustration purposes. For most situations it *may* well be that including "toxicity" as one criterion in the selection of the "worst-case" product may *not* make a difference.

Furthermore, note that my recommendation of performing the protocol on the "most difficult to clean" product but using as the limit the lowest limit of any product in the group would mean for this example evaluating Product A but setting the limit at the limit of Product B. This would provide assurance that Product B, an easier-to-clean product, would be able to be cleaned down to its calculated limit.

Some may object to that recommendation because they would always want to include a highly hazardous product in a protocol (assuming Product B has a highly hazardous active). If that were the case, I would suggest separate validation protocols for Product A and Product B. If there were more than two products, it might be possible to form two groups, one group of highly hazardous actives and one group with actives that were not highly hazardous.

Finally, note that the scheme presented in the "point system" example is just there for illustration purposes. Any point scheme used in a given company should be based on an understanding of products cleaned and an associated risk assessment.

The above chapter is based on a *Cleaning Memo* originally published in November 2017.

35 Another "Worst Case" Product Grouping Idea

I frequently teach that the solubility of the active, when used *alone*, is not necessarily the most appropriate way to select the "hardest to clean" product in a product grouping approach. Note that I sometimes refer to the "hardest to clean" as the "most difficult to clean"; so for purposes of this chapter, consider them interchangeable. Sometimes that approach of only solubility of the active is fine. For example, if I manufacture sterile injectables with all formulation components other than the active (API) readily soluble in water, and if I clean with water alone, it is reasonable to conclude that the product with the active with the lowest solubility in water is the "most difficult to clean". On the other hand, for oral solid dosage forms, such as tablets, it probably is not reliable to depend on the solubility of the active *alone* for determining the "hardest to clean".

Why do I say that? Well, it is a simple factual matter that the excipient components for tablets do affect the difficulty of cleaning. Those excipients may be designed to speed up dissolution (as in orally dispersible tablets) or they may be designed to slow down the release of the active (as in extended release, or ER, tablets). So, the question to answer is "what else can be considered to address those differences?" For example, it may be possible to say that ER tablets are more difficult to clean than immediate release and orally dispersible tablets, and therefore, my first "cut" in a staged approach is to say the ER products in the group are the hardest to clean. Now, I just consider the ER release products, and *among those ER products*, I now select the product with active with the lowest solubility. Will that work? Well, it's a step in the right direction, but how do I know that the excipient formulation among the different ER tablets is equally difficult to clean? Perhaps I might say that the excipient formulation components are essentially the same, and therefore solubility of the active is the deciding parameter. So, I will need to work on those criteria to conclude that the excipient formulations are equally difficult to clean. I suspect you get the idea here of making sure you have a *sound basis* for selection of the "difficult to clean" product in a group.

Okay. Here is a proposal for an alternative method. For oral solid dosage forms like tablets, a common test is a "dissolution time" procedure. This is a test to evaluate the rate and extent of release of the active in tablets. The test apparatus is described in USP <711> [USP, 2016]. The apparatus and test procedure are designed to provide *in vitro* data for the release of active that hopefully correlates with *in vivo release* when taken by a patient. It essentially involves placing a "dissolution medium" in a vessel with a stirrer, placing the product in the dissolution medium, and taking samples at time intervals to measure (by techniques such as

DOI: 10.1201/9781003366003-41

HPLC) the amount of the released active at each time interval. Now, how can this procedure be modified and used for determining the "most difficult to clean" product?

Obviously, the "dissolution medium" is preferably changed to the cleaning solution, which might be water alone or an aqueous formulated detergent formulation at the cleaning temperature. The product is then added to the "dissolution medium" to run the test. Samples are removed at various intervals and analyzed for the active. That data can then be plotted with time on the X-axis and the total amount released on the Y-axis. If two products are compared, then some results can be compared to determine which product is more difficult to clean. If the release is a linear function, then perhaps the rate of release could be used. Another option is to avoid the rate of release issue and just set the criterion as the time to achieve 95% release of the active (or another suitable percentage).

There are some concerns you might already have about testing tablets directly by this method. What if one is a 250 mg gross weight tablet and the other is a 500 mg gross weight tablet? That *might not* be a reliable comparison. Two tablets of identical weight but with different *shapes* (for example, oval vs. spherical) might also be problematic. In those cases, a direct comparison would not be appropriate. I would have to do something different. That might be preparing in the lab tablets of the two different formulations at the same gross weight and with the same physical shape. Or it might not be testing the tablets but testing the powdered formulation *before* the tableting step. After all, when I clean equipment it is not generally the tablet itself that is being cleaned, but some powder from an earlier step or from the (uncoated) tablet itself. Realize here that I might have to develop data comparing the powdered formulation before and after the addition of a lubricant (such as magnesium stearate) to more accurately define the relative differences between different products *at different processing stages* (*if* such differences exist).

Now, let me emphasize this is just an idea. I have never seen this done, nor have I seen it proposed before. So don't adopt this without trying it out and seeing if seems reliable. If this method does, in fact, provide a more reliable and justifiable way to select the most difficult to clean product for a grouping approach, we should share that information.

Now, there may be some objections. One is that rather than use this method, why not just use a simulated cleaning situation in the lab to make the comparison? After all, a simulated cleaning method not only gives the relative difference in the difficulty of cleaning, but also gives, in some sense, an absolute determination of the cleaning conditions (assuming the lab cleaning method reliably translates to a realistic situation on the factory floor). My reply is that "I agree". However, that option has been around for a long time and doesn't seem to be so widely adopted. It may be that a modification of the dissolution testing is something that tablet pharmaceutical manufacturers are familiar with and already have the needed apparatus. Modifying the procedure for a comparison of active release may be more straightforward as compared to designing a lab cleaning method. The second part of my response is that a lab method like the "beaker test" makes a *numerical* comparison difficult, because in a lab cleaning study the evaluation is made after the combination of the washing step *and* the rinsing step, which really adds two elements to the *interpretation* if residues on the test coupons are measured analytically.

A second objection might be that when companies select the "most difficult to clean" product, they also consider the "most toxic" (that is, the active with the lowest limit), so they are not really using solubility of the active *alone*. See Chapter 34 on why toxicity is *not* related to the difficulty of cleaning, but rather to the difficulty of meeting low residue limits in a protocol. That, of course, is why for a product grouping approach I generally recommend selecting the "most difficult to clean" product in the group, but in the protocol evaluating the "most difficult to clean" product at the lowest limit of any product in the group.

Furthermore, you might ask if this type of evaluation could be applied to other formulation types. That's conceivable. For example, it might be able to be adapted for ointments by evaluating a fixed amount (and with fixed shapes) of ointments. It also may apply to liquids if instead of testing the liquid itself, we tested the dried material, which may better simulate the dried material which could be left on the equipment at the end of an extended dirty hold time. However, at this time, the more sensible place to start an evaluation of this technique is with oral solid dosage forms.

This suggestion is not presented as a panacea for all the issues we face in selecting the "most difficult to clean" product in a grouping approach. However, it seems to offer an advantage as compared to just relying on the solubility of the active.

The above chapter is based on a *Cleaning Memo* originally published in June 2020.

REFERENCE

USP <711>. "Dissolution". (2016) https://www.usp.org/sites/default/files/usp/document/harmonization/gen-method/q01_pf_ira_32_2_2006.pdf (accessed May 6, 2021).

Section VII

Protocols and Procedures

The following five chapters cover issues related to the design and execution of cleaning procedures and cleaning validation protocols.

DOI: 10.1201/9781003366003-42

36 Issues in Rinsing
Part 1

In this chapter and in the next chapter, we'll cover issues in the rinse process itself, in rinse sampling, in limits for rinse sampling, and finally in rinse recovery studies. However, before we get to those issues, we'll first need to cover what actually (or hopefully) happens in the *washing step* of a cleaning process (note that I am making a distinction between the washing step, which typically involves the use of cleaning agent such as a detergent, and the rinsing step, which generally involves the use of only a solvent such as water). So, for purposes of this discussion, a *cleaning* process is composed of a *washing* step and a *rinsing* step. Note that a cleaning process may also use a "pre-rinse", which is typically just a solvent (such as water) to physically remove as much "soil" (that is the product you are trying to clean from the equipment) as practical by a mainly physical process. Note further that if you are just cleaning with water alone, it may be more difficult to distinguish what the "break" (or dividing line) might be between the washing step and the rinsing step.

The objective of the wash step is to get the entire product in the equipment after the pre-rinse is dissolved, emulsified, or suspended in the washing solution. Generally, we don't design cleaning processes so that we are leaving a significant amount of product still adhered to the equipment surfaces. Why is that the case? Think of it this way: if I leave a significant amount of product on the equipment surfaces, why am I expecting the rinsing process to remove that product from the equipment surfaces? If the rinse process was that effective, I would just use the rinse itself (for example, water alone) as both my wash solution and my rinse solution [Okay, I know chemically it is not a "solution", but I think you know my intention]. So, in the ideal world, I am using the washing solution to remove everything from the equipment surfaces. If I am not doing that, then I should consider changing my washing solution or my washing solution parameters to make the washing step more effective. For example, I might use a higher wash temperature, a longer wash time, a higher concentration of detergent, or a different detergent. What I think I am trying to convince you (and myself) of is that I want the wash solution to "capture" in some way all product left on the surface (I use the term "capture" to include dissolving, emulsifying, and suspending in the washing solution).

You might ask, how can I determine that occurs except by rinsing the equipment and sampling the surfaces? If that is required, then I am not looking at just the effectiveness of the washing step, but the effectiveness of both the washing step and the rinsing step *in combination*. One way to address only the washing step is to monitor the concentration of the active (assuming my target residue is the

DOI: 10.1201/9781003366003-43

active) in the wash solution as a function of time. This may be possible in a recirculating CIP process using a spray device or in a "fill and dump" process (where there are no spray devices). What I do is sample the wash solution as a function of time, and then measure at each time interval something that is an indication of the degree of removal of the product from the equipment surfaces into the washing solution. For a small molecule active, this might be a simple UV measurement. Yes, it might pick up UV absorbance from the excipients, but that *may* not matter. What I am looking for is an increase in UV absorbance in the wash solution as a function of time until that absorbance levels off, at which point I am no longer removing the product from the surfaces (having captured it in the wash solution). I know you are probably thinking that the reliability of UV absorbance in an emulsified or suspended wash solution might not be so great. In that case, you might have to consider some modification of the sample taken (such as extraction with a solvent) to get a reliable and consistent UV readout. For a biotech active, it might be the use of TOC as a measurement technique. One could expect the TOC value in the wash solution to increase, and then eventually level off when all the organic materials are removed from the surface and captured in the wash solution. Realize in such tests, you may have to use an element of creativity to get reliable measurements. Furthermore, realize that in certain situations, such as in some manual cleaning processes, this technique may not be very practical or very reliable.

Furthermore, realize there may be situations where determining that the measured value in the wash solution *appears* to level off, but is not really an indication that all surfaces are acceptably clean. One case is where there is a dead leg in the equipment. In this situation, you might remove 99.99% of the product, and the response curve appears to be leveling off, but due to the normal analytical method variability, you are still *very slowly increasing* the response. Another case is where there may two different chemical species with significantly different kinetics of solubility in the wash solution. In that situation, you would see an initial leveling off of the response. If you stopped there, you would miss the later jump in response due to the slower dissolving species. It may be the case that when you perform rinsing and examine the equipment either visually or analytically, it would be obvious that something else is going on and that a different approach is needed (which is why these studies are called *experiments*).

The point of discussing the washing process is to set the stage for talking about the rinsing process. If what I say about the wash process is true, and that ideally, the wash solution has *captured* essentially the entire product on the surfaces, what is going on in the rinse process? For this discussion, I could use the example of either a CIP process or a "fill and dump" process. In either case, what is happening is that I am adding more rinse solvent to dilute out the concentration of active (or product) in the wash solution to bring it eventually to an acceptable level. Furthermore, what is left on the equipment is essentially dependent on the amount of final rinse solution left in the equipment and the concentration of the target residue in that final rinse solution. Remember, this is based on the *ideal* situation of all products being captured in the wash solution.

There is another hitch, however. There may be situations where as rinsing continues, I achieve a state where the residue (product) previously captured by the wash solution is no longer still captured. For example, this may happen in cases where the primary cleaning mechanism in the wash step is emulsification. At high concentrations of surfactants in the wash solution, the target residue is trapped in micelles. As rinsing continues and the "available" surfactant concentration drops, the emulsifying ability of the surfactant *may* be lost, and the target residue is no longer emulsified (and therefore no longer captured in the rinse solution). To balance this, it *may* be the case that at very low concentrations the target residue may now be below its solubility limit; the result may be that the residue is now captured by a *dissolving* process. Another example might be a case where the solubility of the target residue depends on the presence of an alkali in the wash solution (here I'm talking *not* about alkaline solutions that degrade the target molecule, but rather a situation where the molecule is intact, but may be more readily soluble at a higher pH). In this situation, as rinsing occurs, it may be the case that as the alkalinity drops, the solubility of the molecule changes significantly. So, absent a significant drop in concentration below the solubility limit at that pH, the target species drop out of the solution. As with the previous example of loss of emulsification, it may be that the resulting concentration is eventually below the solubility limit at the lower pH and will continue to be rinsed appropriately.

So, how does this discussion affect limits for *rinse* sampling? That takes us to the discussion of taking a sample of the *final process rinse* as compared to performing a *separate* sampling rinse (SSR) following the completion of the process rinse [LeBlanc, 2006]. The important point is this. If what was earlier said about the washing step and the rinsing step is true, then what is measured in the final portion of a "fill and dump" process rinse is a direct measure of the residue present on equipment surfaces at the *beginning* of that final "fill and dump" process rinse. Furthermore, what is measured in the *final* portion of a CIP process rinse is likely to be a much lower value as compared to the residue present in the equipment at the *beginning* of that final CIP process rinse. Since what we really want in either situation is a measure of what might be left on equipment surfaces *after completion of the process rinse*, taking rinse samples of the final process rinse represents a worst case.

On the other hand, if we truly want an "accurate" measurement by a rinse procedure of residues on the equipment surfaces *after completion of the process rinse*, an SSR is preferred. In an SSR, a fixed volume of rinse solution is contacted or passed through the equipment, and the *entire* volume is captured to provide a more accurate measurement. In the case of a "fill and dump" rinse, the vessel is filled with rinse solution, agitated for a fixed time, and then any portion of that total volume is analyzed for the presence of residue. For a CIP rinse, a fixed volume of rinse solution is passed through the equipment and the *entire volume* is collected and then agitated for uniformity. A small sample of that uniform solution may then be analyzed for the target residue. The difficulty of a separate collection vessel for the CIP sampling rinse may be avoided by allowing the SSR to collect in the bottom of the vessel, agitating it using appropriate means, and then taking a sub-sample for analysis.

Note that this consideration of washing and rinsing also impacts swab limits. If residues on equipment surfaces at the end of the process rinse are directly related to the amount of rinse solution left in the equipment at the end of the final process rinsing step, then more efficient rinsing of equipment should result in lower carryover. Furthermore, areas where the rinse solution can accumulate and dry are more likely to have higher levels of residue than areas where the rinse solution has effectively drained from equipment surfaces. This suggests that locations where the rinse solution can accumulate after the process rinse are candidates for worst-case swab sampling locations. I realize that in one sense this might appear to fly in the face of the conventional wisdom that worst-case swab locations are those locations based on geometry and design that are more likely to have higher amounts of residues (for example, at stainless steel/gasket junctures) or more difficult to clean residue (for example, at air–liquid interfaces). Rather this analysis suggests that the more important role for these traditional worst-case swab locations is that they should be the locations that drive the *design* of the cleaning process itself. Furthermore, I am not suggesting that these swab sampling locations be discarded. They are still important in confirming that our cleaning process design was, in fact, successful.

If this is an accurate description of what happens in a cleaning process, then the value of designing a washing process that effectively captures all the product provides more assurance that residue levels can be consistently achieved both in validation protocols and in subsequent routine production.

In Chapter 37, we will continue along these same lines and discuss rinse sampling *recovery* studies.

The above chapter is based on a *Cleaning Memo* originally published in November 2019.

REFERENCE

LeBlanc, D.A., "Sampling the Sampling Rinse", in *Cleaning Validation: Practical Compliance Solutions for Pharmaceutical Manufacturing, Volume 1*. Parenteral Drug Association, Bethesda, MD, 2006, pp. 109–110.

37 Issues in Rinsing
Part 2

This chapter is a follow-up to Chapter 36 and focuses on the impact of an understanding of rinse sampling and its effect on rinse sampling *recovery studies*. Please read this only after reviewing Part 1 in the previous chapter. In that first part, we discussed that the objective of the *washing* step in an *ideal world* was to remove all residues from equipment surfaces such that they could be readily removed from the equipment in the *rinsing* step. If that were the case, then the only residue remaining on equipment surfaces would be due to the final rinse solution left in the equipment before drying. But, we also discussed that there may be situations where soils which initially were captured by the washing solution were redeposited during the rinsing step. And, I should add that there probably are situations where we have designed our cleaning processes with the goal of capturing all soils in the washing solution, but we have missed items like dead legs, have misjudged the levels of soil in certain equipment locations, or could not meet acceptance criteria adequately (with a *reasonable* safety margin) with that cleaning process. So, what does this mean for sampling recovery studies?

Typically, sampling recovery studies are designed to *simulate* in some way the rinsing process. They may involve spiking a coupon, suspending it above a clean collection vessel, and then passing a fixed amount of rinse solution across the coupon to collect it in the vessel below. Or, they may involve spiking the bottom of a clean vessel of the appropriate material of construction, adding rinse solution to it, and then mildly agitating it to simulate the time and flow of the final rinse. There certainly are other possibilities. But most involve the application of spiked residues *in solution* to a surface, *drying* the spiked solution on the coupon, and then *approximating* a rinse situation. The question arises "In what way does *drying* the residue on the surface simulate the rinse situation?"

Let's consider the answer to that question in situations where the cleaning process involves a CIP system, and the rinse sample is the final portion of the final *process rinse*. In what sense is doing a recovery study on *dried* residues applicable? The best that could be said is that in most situations a rinse recovery percentage on dried spiked residue represents a *worst case* (a lower recovery percentage) as compared to the actual sampling conditions. I can live with that, perhaps. In addition, since the value of the residue measured in the rinse sample taken in this manner represents a possible worst case as compared to a separate sampling rinse (SSR) taken *after* the process rinse is complete, that provides additional support for doing a rinse recovery using dried residue. If I tried to do a rinse recovery with a spiked coupon that was *not* dried, I am open to the objection that my rinse

DOI: 10.1201/9781003366003-44

recovery procedure might not pick up residue that was *not* adequately captured by the washing solution.

Let's move on to the situation where I'm performing an SSR *after completion of the process rinse*. In this situation, I will generally choose to perform my visual inspection and swab sampling before I perform my SSR. Since ordinarily I prefer to do a visual inspection on dried surfaces and since generally I prefer to do swab sampling on dried surfaces, the equipment will be dried *before* I perform my SSR. Therefore, a rinse sampling recovery on dried spiked residue should be applicable. Let me bring up a few caveats in this particular situation. One is that I want to make sure the swab sampling and/or visual inspection do not interfere with the integrity of that type of rinse sampling. Secondly, if I swab the surface, I will remove residues that could have been picked up by the SSR. However, providing my swab sample passes, this should not affect the validity of the rinse sample (provided that the rinse sample also passes). Assuming the rinse sample limit was based on the same L3 value ($\mu g/cm^2$) as the swab limit, and assuming the swabbed area is small as compared to the rinsed area, it should not affect the validity of a protocol using this approach.

Here is a *second* situation for use of an SSR. What if I choose to *not* perform swab sampling? Could I then choose *not* to dry the equipment before the SSR? In other words, I would complete the process rinse, and then *immediately* begin the SSR. In that situation, could I do a recovery study on residue that was not dried? I would tend to doubt it, for the objection previously discussed (that my rinse recovery procedure might not pick up residue that was not adequately captured by the washing solution) comes into play. Furthermore, the idea of omitting the drying between the process rinse and the SSR causes problems with my visual inspection. If I were to dry the equipment after the SSR and then perform my visual inspection, what have I proved, particularly if my SSR recovery percentage was very high? Of course, even if my cleaning process was not fully effective, I could expect the equipment to be cleaner after that SSR; again, what have I proved? Of course, if the equipment is *not* visually clean after the SSR, I clearly have done something wrong (either in the cleaning process or in the SSR).

Where does this leave us with rinse recovery studies? I wish I could say that rinse recovery studies were not necessary, or that rinse recovery studies could be done on non-dried residues. However, I don't think we are at that point in terms of having data that support those practices, or even in deciding what data should be developed to possibly support those practices. It may be that if we can determine that swab residue values are consistently well below calculated limits in worst-case swab locations, and if we were to include in those worst-case swab locations swab surfaces where a final rinse was likely to pool, perhaps that might help. The complicating factor is that it is hard to generalize from one product to another product, from one cleaning process to another cleaning process, and from one manufacturer to another manufacturer. Furthermore, the way to develop consistent data is certainly more than just three protocol runs (remember the importance of *design and development* in a life cycle approach), and I may not want to omit rinse sampling in the early stages of validation for fear that my swab data alone will not support my desired conclusion. Or, perhaps my objective may not be trying to reduce rinse sampling, but rather trying to eliminate swab sampling.

I fully realize the result of this discussion may not bring us to a clear path forward. However, I hope this two-part series helps us to better understand what is involved in rinsing, rinse sampling, and recovery studies for rinse sampling.

The above chapter is based on a *Cleaning Memo* originally published in December 2019.

38 Routine Monitoring for Highly Hazardous Products

This is a discussion of products with highly hazardous actives and deals with the question of *routine monitoring* for highly hazardous products. I generally define a "highly hazardous product" as a drug product with an active that has a *significant* toxicity concern apart from the therapeutic effect. Others may define it as a drug product containing an active with a PDE/ADE of less than 10 mcg/day. Whichever definition you use (and there is significant overlap), I think you know what I am referring to.

By "routine monitoring" I generally mean testing that is *routinely* done after cleaning validation is done (that is, *after completion* of the validation protocols that are performed as part of the Stage 2 qualification process in a lifecycle validation approach) [FDA, 2011]. By "routinely", I mean testing done either on *every* occurrence of the cleaning process or on some *reduced schedule* such as monthly, quarterly, or based on a frequency such as "every X batches". The testing that might be done includes visual examination, final rinse sample testing, and swab sample testing. *If* visual examination is done for routine monitoring, there *may not* be the same extensive equipment disassembly that is done during the validation protocols. *If* swab sampling is done, there *may* be a reduced number of swab sample locations (selecting, for example, the "worst cases" of what was done in the validation protocols). Finally, the testing that is done *may* be for the drug active, for the cleaning agent, for bioburden, for endotoxin, or for visual assessment. [Note that I have included visual assessment as part of both sampling and analytical methods because it really entails aspects of *both*.]

Let me also clarify that by routine monitoring, I am *not* referring to the practice of repeating one run of a validation protocol on a yearly (or other frequency) basis. However, it is possible to include that practice as part of the Stage 3 "continued process verification" (also called "ongoing process verification" or "validation maintenance") that is a part of a lifecycle validation approach. I'm sure that you will see that some of the things discussed for routine monitoring also apply to that yearly verification protocol or yearly confirmatory protocol. However, the routine monitoring as I usually present it is *not* done as part of a separate protocol, but as part of the cleaning procedure and/or the batch record of that cleaning process.

Okay, I have tried to set the stage. Now let's talk about how this applies to highly hazardous products. As discussed in Chapter 14, those highly hazardous

DOI: 10.1201/9781003366003-45

products are a greater patient safety risk (and consequently a greater business risk), so I want to pay more attention to those products. I certainly should pay more attention to the design/development of the cleaning procedure (lifecycle validation Stage 1) and to the design/execution of the validation protocol (lifecycle validation Stage 2). Our focus *here* is what I should do for routine monitoring (part of lifecycle validation Stage 3). Certainly whatever is done for routine monitoring for non-highly hazardous products should be done *also* for highly hazardous products. But, what *more* should I do?

In most cases, I probably should *not* have to do anything additional for visual examination, cleaning agent testing, bioburden testing, and endotoxin testing. However, it would certainly reduce risk if I were to do some level of additional testing for the highly hazardous active itself. What are the possibilities?

One is to routinely test a final rinse water (or solvent) sample for the active. Rinse sampling (if it encompasses the entire equipment sampled) can be a valuable technique for getting an *overall* picture of any problem that might occur with a cleaning process. Typically, that analysis for highly hazardous products should be a specific analytical procedure, such as HPLC, that tells me *exactly* how much active was present in the rinse sample. I can compare that to both my protocol acceptance limit as well as to trend charts and/or to action/alert levels (the latter having been established based on process capability analysis of rinse sample testing).

It probably is *not* good enough to try to just analyze the final rinse for TOC; TOC can certainly tell you how much carbon is present, but it *cannot* tell you whether that carbon is from the active or from excipients or from processing aids or from cleaning agents. Furthermore, the *practical* LOD/LOQ for TOC samples is probably *not* low enough in most cases for highly hazardous actives to provide a confirmation that a limit is being met for a cleaning validation sample (where you are subtracting out a blank value to determine what could be present due to residues on surfaces). That said, it may be possible *for routine monitoring* to use just a "straight" UV (ultraviolet) method on the rinse sample. You might wonder why I suggest that UV, also a "non-specific" method, might *be* useful, whereas I think TOC is probably *not* useful.

The difference is this. For UV I can generally measure my active at a *much lower concentration* as compared to measurement by TOC. Secondly, there are likely to be *fewer interferences* for the UV measurement as compared to interferences for a TOC measurement. For TOC, most carbon-containing species will interfere and contribute positively to the TOC value. For UV, species (either with or without carbon) will only interfere if they have significant absorption at a wavelength close to the λmax of the highly hazardous active being analyzed. Even if there is some positive interference, you can still show acceptable levels of the highly hazardous active if, based on the total response at the λmax of the highly hazardous active, the assumed value of that highly hazardous active is below the acceptance value (or within trend lines). The value of a straight UV method as compared to an HPLC method may be the *time* required for analysis, and therefore the time required for the *release* of the equipment for subsequent manufacture. Note that even with HPLC analysis only, the equipment may be released *at risk* pending completion of that analysis. However, rather than doing that, I would recommend you perform a UV measurement, and then release it at risk pending a subsequent confirming HPLC analysis.

Enough for rinse sample testing. What about swab sample testing? First, the discussion about selection of an *analytical method* to use for rinse samples *also applies* to swab samples (although for UV analysis, I will also have to consider possible interferences from the swab itself in addition to interferences from the excipients, processing aids, and cleaning agents). Furthermore, as suggested above it may be possible to *reduce* the number of swab samples by only sampling the "worst cases" of the worst cases previously chosen for my validation protocol. In some cases, the determination of the "worst of the worst" may be easy. *For example*, suppose I review the results of the ten sampling locations in a cleaning validation protocol run, and all samples except for two locations gave results below my LOD or below my LOQ. And for those two other locations I consistently (in *all three protocol runs*) obtained results that were acceptable (below my limit), but they were a quantified number above the LOQ. This type of data suggests (but is not an "ironclad" determination) that if I had a problem with my cleaning process, it would show up in those two locations. However, selecting those two locations for swab locations for routine monitoring would be a reasonable approach.

Another possibility for choosing the "worst of the worst" is to look at my initial rationale for selecting swab locations. I may have selected some locations based on complex geometry and other locations based on representative materials of construction. Particularly, if all my protocol data is consistent (that is, I *cannot* separate out those locations with consistently higher values from those with consistently lower values), looking at the initial criterion for the selection of swab sampling locations may be helpful. Let's emphasize, however, that it is *not* necessary to reduce the number of swab sampling locations. However, if a *sound risk assessment* is done, in light of the need to be *more efficient* in manufacturing drug products (without sacrificing quality), the reduction of swab sampling locations for routine monitoring for highly hazardous products should be considered.

Let me close by stating that I am *not* trying to present a "cookie-cutter" approach to routine monitoring for highly hazardous products. What I would like you to do is understand your manufacturing and cleaning processes, and based on a scientifically and logically based risk assessment, decide what is appropriate and applicable for the specific situation in your company.

The above chapter is based on a *Cleaning Memo* originally published in March 2020.

REFERENCES

U.S. Food and Drug Administration, *Process Validation: General Principles and Practices (Revision 1)*, U.S. Government Publishing Office, Washington, D.C., January 2011.

39 "Concurrent Release" for Cleaning Validation

I am generally an advocate of applying the principles of life cycle validation given in the FDA's 2011 process validation [FDA, 2011] to a life cycle approach for cleaning validation (CV). As I have written before, there are many similarities between CV and PV, as well as some significant differences [LeBlanc, 2013a]. One difference I have not written about before is the concept of "concurrent release".

I'll start with a discussion of concurrent release as given in the FDA's PV guidance. Here is the FDA's *definition* of concurrent release as given in that 2011 guidance:

> Releasing for distribution a lot of finished product, manufactured following a qualification protocol, that meets the lot release criteria established in the protocol, but before the entire study protocol has been executed.

The idea here is that *ordinarily* a product is not released *during* the Stage 2 (qualification) protocols. Only after *completion* of the PPQ runs and establishment that a process is validated can the products manufactured during those protocols be released. However, there may be "special situations" where concurrent release of individual batches may be appropriate. There is no need to discuss those situations here (refer to Section V of the FDA PV guidance for that detail), except to note that the PV guidance "expects that concurrent release will be used rarely".

Okay, how do we apply this to CV? The first thing to note is there is a difference between what is released in PV and what is released in CV. In PV, we are releasing the *batches* made in Stage 2 (PPQ) protocols. In CV, we are *not* releasing the batches made, but rather are releasing the *equipment* for subsequent manufacture of the same or a different product. Ordinarily, what happens in CV has no impact on the batch of products just made and cleaned. One exception might be a situation where I obtain non-conforming results in the CV data, and as part of my investigation, I find that there may have been some kind of change (not previously detected) in the manufacturing process of the batch of product just cleaned.

In the discussion that follows, I am assuming I will perform three protocol runs for my cleaning process. So, for equipment release do I have to wait to complete all three runs with acceptable results before I can release any equipment from each of the three runs? Certainly not! That just doesn't make sense, nor is it even possible in most cases. Clearly, the test data on the first protocol run should be adequate to release the equipment for manufacturing of another product (either another batch of the same product or a batch of a different product) on that cleaned

DOI: 10.1201/9781003366003-46

equipment. In this situation, it is not unlike "cleaning verification" [LeBlanc, 2013b], where I prepare a one-off cleaning protocol, such as is commonly done for infrequent production or after a deviation/nonconformance. I then use the protocol data for each of the three runs to release the equipment for another batch or product following each individual run.

Ideally, I should have all the protocol results *for a given protocol run* before I release the equipment for subsequent manufacture. Note that if certain protocol data (such as microbial results) are delayed or not available at the time I would like to use the equipment, I may release the equipment "at risk". In that case, the subsequently manufactured product cannot be considered for product release until that "missing" data has been obtained and the product release has been authorized by QA.

If I am performing three CV runs and the first two have acceptable residue results, but the third has unacceptable residue results, what should be done? Clearly, something may be wrong with the cleaning process. Unless I can somehow invalidate that third run (to make it an *invalid* run, and not a *failed* run), the cleaning process is not validated. If the third run is a *valid failed run*, then I need to make changes as part of my design/development effort and try again with three new protocol runs. However, such a failure of a CV protocol should *not* invalidate the release of the equipment following the first two runs. The data generated in the third run should not change my conclusion regarding the release of the equipment in those two instances. However, for the third run where I obtained unacceptable residue data, clearly, I should clean the equipment again in a "cleaning verification" mode. That cleaning is preferably done with the same cleaning process to avoid the need to calculate residue limits for a different cleaning agent. In most cases where I had such a failed run, repeating the same cleaning process is likely to result in passing data so that the equipment can be released for subsequent manufacture.

In all cases, *product* release should be handled by PV principles and by QA procedures. Except in situations where a failure in a CV protocol suggests a problem in product manufacture or where equipment has been released at risk before all CV analytical data is available, CV results do *not* directly affect what I may be doing for product release.

Note that in some cases, CV may be done on a product where PV was completed many years before. In those cases what is done in CV has no impact on product release (except for the exceptions mentioned in the previous paragraph). And there are some cases where PV and CV are done simultaneously. In that situation, there are three protocol runs for PV, with CV being done by a cleaning process on those *same* batches used for PV. And finally, in some cases, all three PV runs may be completed *immediately before* starting any of the three CV runs. Note that during the PV runs in this third case, there generally will have to be "cleaning verification" performed after each PV run to successfully release the equipment each time for the manufacture of a subsequent batch or product.

A few cautions are in order relating to this issue of "concurrent release" *of equipment*. First is that I have used for illustration purposes that three runs are done for PV and CV. In its process validation guidance, the FDA does not specify the number of runs, while the EU in Annex 15 [European Commission, 2015] still states that the number of batches should be based on QRM principles, but adds that "it is generally considered acceptable that a minimum of three consecutive batches manufactured under routine conditions could constitute a validation of the process". Secondly, there will always be special cases where what I have suggested could be the case probably will *not* be the case. So a careful understanding of what is done in CV, and also *why it is done*, is critical.

The above chapter is based on a *Cleaning Memo* originally published in May 2018.

REFERENCES

European Commission: Directorate-General for Health And Food Safety. EudraLex Volume 4, "EU Guidelines for Good Manufacturing Practice for Medicinal Products for Human and Veterinary Use, Annex 15: Qualification and Validation". Brussels. March 30, 2015.

LeBlanc, D.A., "Differences between Cleaning and Process Validation", in *Cleaning Validation: Practical Compliance Solutions for Pharmaceutical Manufacturing, Volume 3*. Parenteral Drug Association, Bethesda, MD, 2013a, pp. 7–9.

LeBlanc, D.A., "Revisiting 'Cleaning Verification'", in *Cleaning Validation: Practical Compliance Solutions for Pharmaceutical Manufacturing, Volume 3*. Parenteral Drug Association, Bethesda, MD, 2013b, pp. 23–26.

U.S. Food and Drug Administration, *Process Validation: General Principles and Practices (Revision 1)*, U.S. Government Publishing Office, Washington, D.C., January 2011.

40 Dirty and Clean Hold Time Protocols

I sometimes get asked about doing a dirty hold time (DHT) protocol *after* the cleaning validation protocol is complete. Implicit in the question is the assumption that a protocol to determine the effectiveness of the cleaning process is *different* from a protocol to determine the DHT. That is, I first complete my cleaning protocol, and then after the completion of that protocol, I now conduct my DHT protocol.

I can understand why I get that question; however, it *misses the point* about what is being done. Any protocol to evaluate the effectiveness of a cleaning process *always* (I repeat, *always*) has a DHT, even if we choose not to call it a "dirty hold time". In other words, there is always a *finite time* between the end of manufacture (however I define that) and the beginning of the cleaning process (however I define that). It may be only a few hours or a few minutes. But whatever that time is, that is the "validated" maximum DHT for that protocol (just for clarification, that maximum DHT is then the *validated* DHT if the protocol is successful). If I am to use the validated cleaning process in the future, I must initiate cleaning *within* that maximum DHT, or else I face a deviation (or nonconformance if you prefer that term). There is no way to get away from this situation; there is *always* an *inherent* DHT in a validation protocol for a cleaning process.

Now, after performing a cleaning validation protocol with a specified DHT, it may be possible to *increase* that maximum DHT by performing another protocol. However, unless other things are also changing, that new protocol should have the same conditions and acceptance criteria as the previous protocol (except, of course, for the longer DHT). The terminology I would use to refer to this new protocol is not "I am validating a longer dirty hold time", but rather "I am validating a cleaning process with a longer dirty hold time" (although I think understand what is being said by the former phrase). The reason for this is that a DHT is a parameter (a variable one) in the cleaning process. It is not unlike the washing time, the washing temperature, and the cleaning agent concentration. It is a variable or parameter to be controlled for the cleaning process. If I extend the DHT, I am therefore validating a cleaning process with that (new) parameter.

In contrast, when I talk about the clean hold time (CHT), I generally refer to a CHT protocol. Perhaps I am being somewhat inconsistent, but the objective of the CHT is to establish that the equipment does not become recontaminated between the *end* of cleaning and the *beginning* of manufacture of the next product (or the beginning of a SIP process). I could say I am validating the process for "storing" the equipment in between uses, but I don't do anything other than test it at the end

DOI: 10.1201/9781003366003-47

of the storage time under defined storage conditions (and perhaps compare that data to the "beginning of storage" data). That is, any actions I take to protect the equipment during "storage" or "idle time" ideally should be part of the cleaning SOP (although if you pushed me, I would say it is okay to have one SOP for the cleaning process and a second SOP for preparation for storage under the CHT).

Furthermore, the CHT protocol can be a *separate* protocol from the cleaning process protocol. That is, it is possible (and is my preference, because I am "hardwired" to like modularity) to have one protocol that determines the effectiveness of my cleaning process under defined conditions, including a maximum DHT. Then I can perform a second protocol for the CHT, to evaluate the maintenance of a clean state during the CHT. This second (separate) protocol can be done "back-to-back" with the cleaning process protocol. In that case, the microbial data from the end of the cleaning protocol may serve as the "time zero" or baseline microbial data for my CHT protocol. Or, I could separate these two protocols and perform the CHT following a cleaning event that was *not* part of the cleaning process protocol. The reason I like this "distinction" between the two protocols is that I only execute the CHT protocol if the cleaning process protocol is successful. If the cleaning process is not validated, what value is it to conduct the CHT protocol (although there may be unusual circumstances where performing the CHT part of the protocol could appropriately follow an *unsuccessful* cleaning process protocol)?

Needless to say, it is still acceptable (and probably more common) to write one protocol with instructions and acceptance criteria for *both* the acceptability of the cleaning process *and* the acceptability of the CHT storage parameters. However, in such situations, I should be prepared to deal with situations where the cleaning process part of the protocol is successful, but the CHT part of the protocol fails.

For clarification, when I refer to the storage conditions as part of the CHT protocol, I am not referring to storage in an uncontrolled area or storage for an extended time (such as months). I am assuming the equipment is used regularly, and the storage conditions are the "normal" conditions between manufacturing uses of that equipment.

The above chapter is based on a *Cleaning Memo* originally published in April 2017.

Section VIII

API Manufacture

The following three chapters deal with cleaning validation in API manufacturing situations, particularly in small molecule API synthesis.

DOI: 10.1201/9781003366003-48

41 A Critique of the APIC Guideline

The APIC "Guidance on Aspects of Cleaning Validation in Active Pharmaceutical Ingredient Plants" was revised in 2016 [APIC, 2016]. This guide is one of the few that specifically addresses issues with cleaning validation in *small molecule API* facilities (those that manufacture APIs by chemical organic synthesis, usually in solvents). While some of the concepts may apply to bulk biotechnology manufacture, the guide does not seem to be oriented to that type of manufacture.

The focus of the 2016 revision is to incorporate the terminology of PDE (Permitted Daily Exposure) from the EMA health-based limits *draft* guideline [EMA, 2014], in addition to the ADE (Acceptable Daily Exposure) terminology which was used in the prior (2014) version of the APIC guideline [APIC, 2014]. Most other things survive from the 2014 version. However, I have not critiqued the 2014 version in a Cleaning Memo, so I will take this opportunity to do so now with this updated version.

First, the positive. This guide (as in the past) focuses on what is different and unique in small molecule API synthesis. The issues related to how to handle intermediates and how to handle "clearance" are a highlight of this guidance. Yes, ICH Q7 says a little about these issues but provides little specific guidance [ICH, 2000]. If you manufacture small molecule APIs and you have not read this or an earlier version of this APIC guide, you should do so now (on the other hand, if you are not familiar with this guide, this Cleaning Memo will probably not make sense to you).

My critiques are for the most part relatively minor and are basically there to further clarify some issues that are not explained fully (but which should be considered in applying this guide, or considered in future revisions of this guide). I will refer to them by section number in the 2016 guidance.

For clarification, the term MACO as used by APIC is equivalent to the expression L2 that I typically use. The term MAXCONC is the concentration of the next product (L1) limit that I typically use.

DAILY DOSE DEFINITION

In the MACO calculation based on the therapeutic daily dose (in Section 4.2.2), the "Standard Therapeutic Daily Dose" for both the cleaned product (in the numerator) and for the next product (in the denominator) is used. Typically for carryover calculations, the *minimum* therapeutic daily dose is used for the *cleaned* product and the *maximum* therapeutic daily dose is used for the *next* product. It is unclear why this minimum/maximum is not used here. This same comment

DOI: 10.1201/9781003366003-49

applies to the health-based calculation in 4.2.1 and the LD50 calculation in section 4.2.3, although in these cases it is only the terminology (or definition) of the daily dose of the *next* product (in the denominator) that is relevant.

SAFETY FACTORS FOR LD50 CALCULATIONS

In section 4.2.3, the identification of the "way of entry (IV, oral, etc.)" for the LD50 testing is specified. Then an empirical factor (2000) and a safety factor are applied to that LD50 to take it down to a safe level. Those safety factors, which depend on the route of administration of the drug, are as follows:

Topicals	10–100
Orals	100–1,000
Parenterals	1,000–100,000

While it is certainly possible to use different safety factors for different routes of administration, the critical element is whether the LD50 route matches the route of administration of the *next* product. This may be implicit in this section, but it should be *explicit*. In other words, if my LD50 of the cleaned product is oral and the next product is a parenteral, my safety factor should be greater as compared to a case where the cleaned product is orally administered and the next product is also orally administered.

The *same* routes of administration for what is in the numerator and in the denominator of a MACO calculation are also relevant to the *dose-based* calculation in Section 4.2.2. If the cleaned product is an oral dose and the next product is a parenteral, I probably need an additional safety factor. Yes, the guide states that 1,000 is "normally" used, but it would be beneficial to state when it is *not* used. Unlike the situation in a *finished drug product* facility where it is likely that all products made in a given facility are all by the same route of administration, in many cases, for *small molecule API* manufacturing, some APIs are for oral dosing and some for parenteral dosing.

Finally, this comment does not apply to the health-based limit MACO in 4.2.1. By the traditional definitions, ADE/ PDE values are by *any* (meaning all?) route of administration. In this case, the ADE/PDE is based on parenteral use (more or less 100% available systemically). Unless the ADE or PDE is for a defined specific route of administration [LeBlanc, 2017], then this concern is not relevant.

GENERAL LIMIT

There are two minor but possibly confusing things in Section 4.2.4. First is the use of the term "MACOppm". What this means (I think) is the MACO calculated on a MAXCONC value in ppm. My first reading of this was that the MACO was expressed as ppm, which is not the case since MACO is in mass units such as mg. While the APIC authors used the same "MACO" for MACO calculated by either the health, dose, or LD50 calculations, it seems unnecessary (and possibly) confusing to add the "ppm" to the acronym for the use here. It also seems confusing because MACOppm is used in the equations, but just the term MACO (without the "ppm") is used in the example given following the equation.

In this same section, MAXCONC is the maximum concentration in the next product. Concentration is given as "kg/kg or ppm". It is unclear whether this is an option (you may express it as either kg/kg or as ppm), or whether this is a typo and that it should read "mg/kg or ppm". Either is possible. My guess is that it is a typo, because if the MAXCONC is in kg/kg, then either MACO is expressed as kg (resulting in a value with lots of decimal places), or else an additional conversion factor has to be used to convert that kg to something smaller, like μg or g or mg. Furthermore, in the definition of MACOppm, it is stated that it is "Calculated from general ppm limit", which seems to imply that the MAXCONC units would be ppm (or mg/kg).

In this same section is Equation 4.2.5-II, giving a value for "CO", the "True (measured) total quantity" calculated from the swab results. (I think CO refers to the "CarryOver"). However, I'm unclear what the distinction is between CO used here and M used in Section 4.2.6, except that one refers to the total actual carryover for swab sampling and the other refers to total actual carryover for rinse limits. The idea in this equation is that based on actual swab results (measured in a protocol, for example), it is possible to calculate the total carryover and compare that value to the total carryover allowed (the MACO). While there are certain circumstances where that is possible, the conditions under which it can be used are not given. This equation can only be used if I assume an approach much like that described in situations related to "stratified sampling" [LeBlanc, 2013a; LeBlanc, 2013b; LeBlanc, 2013c]. If I only take one swab sample for a given vessel and multiply that actual value in μg/dm^2 by the surface area of the vessel (in dm^2), I can get a value for the total carryover in μg. This assumes, of course, that I have selected the worst-case swabbing location. However, it is more likely to be the case that I would choose multiple swabbing locations as worst cases. Unless I multiply only the highest swab value by the total equipment surface area, or else use a stratified approach, I may understate the total carryover.

Why this equation (and calculation) is confusing is that the calculation gives a summation of different swab values. This summation is correct if it involves, for example, a series of different pieces of equipment, and I multiply the swab results by the area of that specific piece of equipment, and then I add the values for different pieces of equipment together. Where I have seen this misused is where within one piece of equipment, the different swab locations are multiplied by the area sampled, not by what is given in the guide as Ai, or "Area for the tested piece of equipment # i". Ai is to be the *total equipment area*, not the specific area sampled for that swab result.

If there are multiple swab locations for one piece of equipment, and if those specific swab locations can be linked to a certain *segment* of the overall area of that piece of equipment (use of the "stratified sampling" approach within a given equipment item), then it may be possible to use that approach when there are multiple swabbed locations for a given piece of equipment [LeBlanc, 2013a].

A final minor point is that the definition of "mi" is given as "quantity in µg/dm^2". This could be better expressed as "value in µg/dm^2". When I hear the word "quantity", I usually think of an amount in mass (for example), not a concentration or an amount per surface area. I may be being overly picky for this last point, however.

RINSE LIMIT

In section 4.2.6, equations are given for determining the acceptability of the rinse limit. The requirement is given as "M < Target value". As defined M is an "amount of residue in the cleaned equipment in mg" and the "Target value" is a concentration (in mg/L) in the rinse sample. Obviously, it is *not* appropriate to compare one value in mg to another with units of mg/L. I suspect what is intended is that "Requirement: M < MACO". That way, both values are amounts in mg.

CONFUSING TERMINOLOGY

In Section 8.0 (on "Determination of the Amount of Residue", relating to analytical methods used), the term "Mper" is introduced and used several times without clearly identifying what it refers to. The first use is in 8.1 where the "carryover acceptance limit" is referred to as Mper. I think, but it is not stated, that Mper refers to the *permitted* value of M. That same section then refers to M as the "actual amount of residue" left in the equipment (which is consistent with the use of M in section 4.2.6).

Section 8.2.4 introduces the term Mres, which is the *measured* amount of residue by the analytical method, which is then divided by the recovery (expressed as decimal) to give M, which is the *true* amount of residue. I suspect that the "res" suffix for M is derived from "result" (meaning the analytical result), and not "residue" (my original *incorrect* assumption).

Section 8.2.5, dealing with *analytical method validation* for a range, gives MperMin (the lowest foreseen acceptance limit) and MperMax the "highest limit". However, while validating an analytical method from the upper permitted limit down to the lower permitted limit is acceptable, what this actually means is that for the situation where I am measuring residue in the *minimum* situation, I really just have a pass/fail test. Most companies in that situation would want to extend the validated range significantly *below* the MperMin in order to demonstrate the robustness of their cleaning procedure. For example, if the minimum *permitted* M is X, I would prefer to have an analytical method that could measure down to situations where the *measured* M was closer to 0.1X.

SWAB SAMPLING EQUATION

Equation 6 in Section 8.3.1 gives an equation for calculating the total "true" carryover (M) based on averaging the individual results of each swab sample (in units such as mg/dm^2), and then multiplying that value by the total equipment surface area and adjusting for the sampling recovery factor (or recovery rate). This is *not* acceptable *unless* a stratified approach is used (see the earlier discussion in "General Limit"). For example, if I could use this equation, I would just sample a large number of easy-to-clean locations, with the result that any high values in worst-case locations would be effectively averaged down to an acceptable value. Also, when I measure *more than one* worst-case location in a given piece of equipment, how do I know what part of the total surface area each worst-case sampled area actually represents? As I stated earlier, unless the highest value of an equipment swab sample is used for the total area, or else stratified sampling is used for segments within a given equipment item, this Equation 6 should be avoided.

While this guide is valuable and is the best available for small molecule API synthesis, use it fully understanding how it applies to your specific situation.

The above chapter is based on a *Cleaning Memo* originally published in January 2017.

REFERENCES

Active Pharmaceutical Ingredients Committee (APIC). "Guidance on Aspects of Cleaning Validation In Active Pharmaceutical Ingredient Plants". Revision September 2014.

Active Pharmaceutical Ingredients Committee (APIC). "Guidance on Aspects of Cleaning Validation in Active Pharmaceutical Ingredient Plants". Revision September 2016.

European Medicines Agency. "Guideline on Setting Health Based Exposure Limits for Use in Risk Identification in the Manufacture of Different Medicinal Products in Shared facilities". Document EMA/CHMP/ CVMP/SWP/169430/2012. London. November 20, 2014.

ICH Q7, "Good Manufacturing Practice Guide for Active Pharmaceutical Ingredients". 10 November 2000.

LeBlanc, D.A., "Basics of 'Stratified Sampling'", in *Cleaning Validation: Practical Compliance Solutions for Pharmaceutical Manufacturing, Volume 3*. Parenteral Drug Association, Bethesda, MD, 2013a, pp. 93–96.

LeBlanc, D.A., "More on 'Stratified Sampling'", in *Cleaning Validation: Practical Compliance Solutions for Pharmaceutical Manufacturing, Volume 3*. Parenteral Drug Association, Bethesda, MD, 2013b, pp. 97–100.

LeBlanc, D.A., "Final Notes on 'Stratified Sampling'", in *Cleaning Validation: Practical Compliance Solutions for Pharmaceutical Manufacturing, Volume 3*. Parenteral Drug Association, Bethesda, MD, 2013c, pp. 101–104.

LeBlanc, D.A., "Route Specific Health-based Limit Values", in *Cleaning Validation: Practical Compliance Solutions for Pharmaceutical Manufacturing, Volume 4*. Parenteral Drug Association, Bethesda, MD, 2017, pp. 29–32.

42 Another Issue for API Synthesis

I critiqued the APIC cleaning validation guide for small molecule API (drug substance) manufacture in Chapter 41 of this volume. That APIC guide [APIC, 2016] has many good points and a few areas for improvement. One issue not addressed by that guide is dealing with the processing of multiple lots of the *same* product (by "product" I mean either the synthesized intermediate or the synthesized API itself). The schematic in the APIC guide (Figure 1 on page 20) covers the appropriate approach where in the same equipment one is going from one intermediate to another intermediate or the final API in the *same* synthesis chain, from one intermediate to an intermediate or final API in a *different* synthesis chain, and from one final API to either a *different* final API or to *earlier* intermediate in a *different* synthesis chain.

The focus of this chapter is on what is required (or recommended) if I go from one product (either an intermediate or an API) to a second batch or a lot of the *same* product in the *same* equipment. That is, what should be considered in this type of risk assessment? Unless there are some unusual issues, this situation of going from one product to another batch/lot of the same product should represent a *lower* risk as compared to going from one product to a different product. I might ask, what is the risk if I leave a small amount of the previous product behind on equipment surfaces, and the next product made in that equipment is another batch of that *same* product? Clearly, there may be a lot of integrity issues, and I may choose to have more robust cleaning to minimize concerns over batch comingling.

But, another concern is whether the residues from one batch *interfere* with the synthesis in the processing of the next batch of that same product. Suppose my reaction is as follows:

ReactantA + ReactantB → ProductC

Does the presence of ProductC at the beginning of the next batch synthesis (where I am adding ReactantA and ReactantB) cause any *side reactions* to occur? In many cases, the answer is probably "No", because as soon as I start my synthesis of that next batch, ProductC will be present.

However, there may be situations where what is happening in the synthesis of a product is slightly different. Let's suppose that the process involves two reactants to form a given product, and then without isolation of that product, I then add another reactant to form ProductT. This type of synthesis is presented below.

Synthesis Step #1: ReactantQ + ReactantR → IntermediateQR

DOI: 10.1201/9781003366003-50

Synthesis Step # 2: IntermediateQR + Reactants → ProductT

In this case, ProductT is not ordinarily present in significant quantities during the reaction of the first two reactants, and conceivably it could interfere with that initial reaction. This may require additional evaluation to determine whether significant quantities of ProductT left in the reactor would cause unacceptable side reactions (another way to look at this is to determine how much of ProductT can be left in the reactor without causing unacceptable side reactions in the subsequent initial synthesis).

Does this mean, particularly in the earlier example, that I don't have to clean at all? Perhaps, but I may want to do something to minimize *gross* batch intermingling. I may just do a solvent flush, perhaps just using the solvent that is used for that next synthesis step. This does not have to be a "full" cleaning process, nor is it necessarily required that the equipment be visually clean. In such a case, I would not even call this a cleaning step, but would use terminology such as a "solvent flush" or "process solvent rinse".

An analogy is what is done in drug product tablet manufacture for multiple batches (such as eight batches) of the *same* product. I may process eight batches of the same formulation on my tablet press, and only perform a validated cleaning process at the end of those eight batches. In between any two batches, I will just vacuum or brush the equipment to remove gross amounts of powder. Such vacuuming or brushing may not result in equipment that is visually clean. However, it prevents the buildup of residues which could affect batch comingling, could affect the *functionality* of the equipment in the tablet processing, and ultimately could affect the product *quality* of a batch near the end of a campaign.

In the same way, I could use a solvent flush between batches of the same product in small molecule synthesis to minimize comingling of lots and to minimize the possibility of any unacceptable side reactions from occurring. Realize, however, that this analogy is not a perfect analogy. In the tablet manufacture, there is generally not a concern about the residue of the same product reacting with, and changing in some chemical way, the next batch. With small molecule organic synthesis, the effect of the residue on the synthesis itself may, in some circumstances as mentioned earlier, be significant.

Clearly, if this "solvent flush" approach is considered, I would still like to compare product *quality* for successive batches of the same product with only a solvent flush between batches. One thing I would look for is any change in the *impurity profile*, particularly looking for any trends and/or the presence of any unknown peaks in the HPLC analysis. Note that I *may* see differences in comparing impurity profiles of "Batch 1" to "Batch 2" (in particular I might find that the second batch is "cleaner"). Why is that the case? It may be the case because "Batch 1" is likely to see some residues from the prior cleaning process of a different product, while residues from that "before Batch 1 cleaning process" are not likely to be a significant issue for Batch 2, Batch 3, and so on.

A second concern relating to the impurity profile is whether I see a change in the impurity profile from Batch 1 to subsequent batches of the same product due to the fact that degradation occurs during the time interval between batches due to such factors as heat, light, and oxygen. While degradation might also occur if the equipment is more rigorously cleaned between batches, the level of degradants in that situation would likely be significantly lower as compared to just using a solvent flush between batches.

Note that any addition of a solvent or process material *after completion* of the initial synthesis reaction may affect how this is handled. For example, if a different process solvent is added to precipitate the reaction product, the presence of that second solvent at the *beginning* of the next synthesis step of the next batch may be problematic. If it does cause concerns, then a process solvent rinse (that is, a rinse with the solvent used at the beginning of the next batch) would definitely be needed.

Clearly, at the end of a series of batches of the same product, I would want to clean more rigorously. The extent of cleaning in that situation would depend on the product or products made in that next synthesis step. Both effects of residues of the cleaned product on that next synthesis itself and any health/safety issues if those residues transfer to the final API should be considered.

The issues discussed in this chapter focus on *batch-to-batch* residue concerns in small molecule organic synthesis. Needless to say, this discussion illustrates the importance of issues other than patient safety being important for cleaning validation.

The above chapter is based on a *Cleaning Memo* originally published in September 2017.

REFERENCES

Active Pharmaceutical Ingredients Committee (APIC). "Guidance on Aspects of Cleaning Validation in Active Pharmaceutical Ingredient Plants". Revision September 2016.

43 Contaminants in API Manufacture

Starting about 2018, the FDA (as well as many other regulatory agencies) has been dealing with recalls of drug products and drug substances contaminated with nitrosamines [FDA, 2021a]. Nitrosamines are probable human carcinogens. The products are associated with API manufacture of angiotensin II receptor blockers (ARBs), mainly with valsartan but also including other ARBs. The two nitrosamine impurities that are generally found are N-Nitrosodimethylamine (NDMA) and N-Nitrosodiethylamine (NDEA). Those two materials are found widely in water and foods, but the FDA has stated that their "presence in drug products is unacceptable".

The safe daily intake amount for the suspect nitrosamines is 0.096 µg/day. The amounts found in *some* tablets of ARB products have ranged from 0.3 µg/tablet up to 20 µg/tablet, clearly above the safe threshold set by the FDA. The FDA has been clear to point out that not all lots of ARB products have unacceptable levels. The FDA has published analytical methods suitable for measuring NDMA and NDEA in valsartan drug products and drug substances, which it believes would be suitable for other ARB products.

The FDA as well as EU agencies are trying to determine the *root causes* of the nitrosamine impurities, as well as to *prevent the recurrence* of these problems. The FDA describes its effort as an "exhaustive effort". Two possible root causes being investigated are (a) by-products of the synthesis procedure and (b) reuse of solvents. And the two may be related.

Nitrosamines are generally formed by the reaction of sodium nitrite with secondary amines. While neither is required in the synthesis of valsartan itself, in a *newer* synthetic route, sodium nitrite is used to destroy the excess sodium azide used in that newer process. If the DMF (dimethylformamide), which is the solvent used in the ARB synthesis, contains residual dimethylamine, this could lead to the formation of NDMA (a common synthesis of DMF uses dimethylamine as a reactant).

So, why is this important for the *cleaning validation community*? Well, it reminded me of two past problems with API syntheses. The first is the case mentioned in the 1993 FDA Cleaning Validation guideline [FDA, 1993]. It involves a drug product made with a drug active cross-contaminated with an intermediate and/or degradant from the synthesis of an agricultural pesticide product. The cause appeared to be reclaiming and reuse of a solvent without making sure the solvent quality was not inappropriately compromised with residues that could be transferred to the drug substance. This was the event cited by the FDA that increased its awareness of the need for validating cleaning processes.

The second problem was with Roche's Viracept (over ten years ago) [Killilea, 2012]. In that situation, ethanol was used to clean a storage tank for methane sulfonic acid (MSA), and was apparently not adequately removed (or evaporated) prior to the addition of the MSA reactant. Over time, the ethanol and MSA *slowly* reacted to form ethyl mesylate, a known carcinogen. The reaction of MSA with nelfinavir to produce the desired API (nelfinavir mesylate) occurred in an ethanol solvent, but because of the *short* reaction time, the production of ethyl mesylate as a by-product in the reaction step itself was minimal. Only with the continued use of the MSA reactant (now with higher levels of the carcinogenic impurity) did the problem of ethyl mesylate in the API become apparent.

So, either of these two older situations may shed light on possible root causes for the more recent nitrosamine contamination issues with ARBs. And they may point out that prior acceptable limits for "impurities" may not be applicable in all situations, that closer attention may be required to possible by-product reactions to form highly toxic impurities, and that closer scrutiny may be required for the quality of solvents (whether a "virgin" solvent or a reclaimed solvent).

There are various subsequent updates from the FDA and other regulatory agencies that should be consulted for issues specifically related to nitrosamine impurities in drug products [EMA, 2021; FDA, 2021b].

The above chapter is based on a *Cleaning Memo* originally published in March 2019.

REFERENCES

EMA. "European Medicines Regulatory Network Approach for the Implementation of the CHMP Opinion Pursuant to Article 5(3) of Regulation (EC) No 726/2004 for Nitrosamine Impurities in Human Medicines". EMA/425645/2020. 22 February 2021. https://www.ema.europa.eu/en/documents/referral/european-medicines-regulatory-network-approach-implementation-chmp-opinion-pursuant-article-53/2004-nitrosamine-impurities-human-medicines_en.pdf (accessed May 6, 2021).

FDA, "Guide to Inspections Validation of Cleaning Processes". United States Printing Office, 1993.

FDA, "Information about Nitrosamine Impurities in Medications". 24 February 2021a https://www.fda.gov/drugs/drug-safety-and-availability/information-about-nitrosamine-impurities-medications (accessed May 6, 2021).

FDA. "Control of Nitrosamine Impurities in Human Drugs: Guidance for Industry". Revision 1 February 2021b. https://www.fda.gov/media/141720/download (accessed May 6, 2021).

Killilea, M.C. "Cleaning Validation: Viracept, 2007". *Journal of Validation Technology* 2012, 18(4), pp. 12–14, November.

Section IX

Miscellaneous Topics

The following five chapters deal with miscellaneous issues not clearly fitting in one of the other major groupings of topics.

DOI: 10.1201/9781003366003-52

44 Significant Figures
Back to Basics

Those of you that have taken any of my training courses or webinars know that I have certain "pet peeves". One of them is *significant figures*, particularly in expressing sampling recovery factors. I commonly see in company reports and in published papers that recovery percentages are given to three or four decimal places. For example, a swab recovery study result is reported as 73.468%. Well, swabbing is like a type of manual cleaning, and the numbers are just *not* that precise. There are several elements that contribute to this misuse. One is that often the input data is just two or three significant figures. So if the measured amount per swab in the recovery sample is 127 μg and the amount spiked is 150 μg, the recovery is reported as 84.667%. And then if there are three replicates, for example, with recoveries of 84.667%, 78.667%, and 87.333%, the average is reported as 85.556%. Another reason good scientists report numbers that way is probably because they use an Excel spreadsheet for calculations, and the spreadsheet is set up to automatically report to three (or four) decimal places.

According to the well-accepted rules involving significant figures, if the input data is only three significant figures, the reported result of the calculation should be rounded to three significant figures. Now there is an exception to this in that if the result of a calculation is used in subsequent calculations, the unrounded result may be used in those subsequent calculations, but the *final* result should be reported to the appropriate number of significant figures. Furthermore, it is generally accepted that while the unrounded number may be used for subsequent calculations, what is *reported* should be reported with the proper number of significant figures. I believe the reason for this is that if a number is used for a large number of subsequent calculations (for example, ten subsequent calculations) and if rounding is done at *each* calculation, the final result conceivably could be "off" by a digit or more in the final result (at the end of the tenth calculation.)

A second, and what I consider a more significant, reason for not reporting a sampling recovery percentage to three or four decimal places is that regardless of the number of significant figures in the input data used to calculate the percentage, the resulting percentages, as a *practical* matter, are just *not that precise*. In the example given previously with percentage values from three replicates, the best that I could have any confidence in would be two significant figures (that is, with no digits to the right of the decimal). So I would report the average (my "official" recovery percentage) as 86%. If the "official" recovery percentage was to the lowest of the three replicates, then that would be 79%.

Some might object that rounding up (that is rounding 85.556% up to 86%) makes it easier to pass in an executed protocol, and that we should really be using

DOI: 10.1201/9781003366003-53

the lower value of 85.556%, or even use "rounding down" principles to use 85%. My belief is that if the difference between 86% and 85% recovery is the difference between passing and failing in a protocol, then I should really design my cleaning process to be more robust.

Some might further argue that this discussion is like asking "how many angels can fit on the end of a pen?" If I have a robust cleaning process, it doesn't matter how many significant figures I report. While there is some truth to that, it does reflect a "casualness" to well-established scientific practices.

There are some good sources for us in the pharmaceutical industry to utilize in addressing significant figures and rounding in the use and reporting of data. One is Section 4 of Volume III of the FDA's "ORA Laboratory Manual" on statistics and data presentation [FDA, 2019]. Here is what that manual says about significant figures:

Definitions and Rules for Significant Figures
- All non-zero digits are significant.
- The most significant digit in a reported result is the left-most non-zero digit: 359.741 (3 is the most significant digit).
- If there is a decimal point, the least significant digit in a reported result is the rightmost digit (whether zero or not): 359.741 (1 is the least significant digit). If there is no decimal point present, the right-most non-zero digit is the least significant digit.
- The number of digits between and including the most and least significant digit is the number of significant digits in the result: 359.741 (there are six significant digits).

And here is what the FDA manual says about significant figures in *calculated* results:

Most analytical results in ORA laboratories are obtained by arithmetic combinations of numbers: addition, subtraction, multiplication, and division. The proper number of digits used to express the result can be easily obtained in all cases by remembering the principle stated above: numerical results are reported with a precision near that of the least precise numerical measurement used to generate the number. Some guidelines and examples follow.

Addition and Subtraction

The general guideline when adding and subtracting numbers is that the answer should have decimal places equal to that of the component with the least number of decimal places:

21.1
2.037
<u>6.13 </u>
29.267 = 29.3, since component 21.1 has the *least* number of decimal places

Multiplication and Division

The general guideline is that the answer has the same number of significant figures as the number with the fewest significant figures:

$$\frac{56 \times 0.003462 \times 43.72}{1.684}$$

A calculator yields an answer of 4.975740998 = 5.0, since one of the measurements has only two significant figures.

And here is what the FDA manual says about rounding:

The following rules should be used:

- If the extra digit is less than 5, drop the digit.
- If the extra digit is greater than 5, drop it and increase the previous digit by one.
- If the extra digit is five, then increase the previous digit by one if it is odd; otherwise do not change the previous digit.

Another good source is the USP General Notices on "Rounding Rules" [USP 38, 2021]. It differs slightly from the FDA approach for rounding in that the third and fourth bullet points from the FDA would be replaced with something like "If the extra digit is *5 or greater than 5*, drop it and *increase* the previous digit by one".

What is discussed here probably has no significant effect on the outcome of a protocol. However, if we become sloppy using one well-established scientific practice, it is likely in the future that this sloppiness could carry over to other well-established practices.

The above chapter is based on a *Cleaning Memo* originally published in February 2019.

REFERENCES

FDA. "ORA Laboratory Manual Volume III Section 4: Basic Statistics and Data Presentation". 13 August 2019. https://www.fda.gov/media/73535/download (accessed May 6, 2021).
USP 38. "General Notices and Requirements". https://www.uspnf.com/sites/default/files/usp_pdf/EN/USPNF/usp-nf-notices/usp38_nf33_gn.pdf (accessed May 6, 2021).

45 The Value of a Protocol Worksheet for Manual Cleaning

In a protocol for pharmaceutical cleaning validation, there are generally many worksheets associated with that protocol. For example, there might be a worksheet for entering sampling locations and ID numbers, with the time and date of sampling, along with the initials of the sampler. There might be a similar worksheet for analytical results, including analytical results for actives and cleaning agents, microbial results for bioburden, and results for visual examination. However, one worksheet that is particularly important for *manual cleaning* is a worksheet for the *execution* of the manual cleaning process. This worksheet is *an independent verification* that the manual cleaning process was performed correctly.

The worksheet would be completed not by the cleaning operator, but by a second person (perhaps from a validation or quality assurance department) who observes the manual cleaning process and the operator from the beginning to the end. That observer uses a worksheet with a list of specific actions or steps to be taken by the cleaning operator. That list of specific actions should closely follow the specific actions given in the SOP of the cleaning process to be validated. As each action is done correctly, the observer checks off that the action was completed correctly (or not), perhaps giving any additional comments as to what was not done correctly or possible ways to improve the written cleaning SOP.

You might ask why an independent verification is necessary, since the cleaning operator himself or herself documents that the cleaning process was done correctly in the cleaning log. Shouldn't the cleaning log be enough? Well, let's step back and remember that we are talking about a *manual* cleaning process. An independent verification has value in several situations in a protocol. For review, the purpose of the protocol is to establish that *if the cleaning process is done correctly*, I will obtain acceptable residue results. What happens in a manual cleaning process if I don't have an independent verification that the cleaning process was done correctly and I end up with *failing* residue results? Where do I look for the root cause of the failure? Was it the analytical group who did something wrong in the lab? Was it the sampler who made a mistake? Did the cleaning operator not follow the SOP? Or, was the cleaning process just a poorly designed cleaning process? All those are possibilities, but inevitably a sharp eye is usually cast on the cleaning operator. Besides the operator saying "I followed the SOP correctly", how can that last concern be investigated?

DOI: 10.1201/9781003366003-54

Well, one way is to have an independent verification to confirm whether the cleaning process was done correctly or not. In a sense, it protects both the integrity of the protocol and the *integrity of the cleaning operator*. If there are failing results and the observer confirms that the operator did everything correctly according to the SOP, then that should eliminate the operator as the root cause. The investigation should perhaps focus elsewhere, including the design of the cleaning process. If there are failing results and the observer notes where the operator performed a step incorrectly or omitted a step, then that helps identify a possible attributable cause, and the cleaning protocol run can be considered not a failed run, but rather an *invalid* run. Of course, the fact that it is invalid may be small consolation, because that operator (and perhaps all operators) has to be trained again with an emphasis on what went wrong; then a new protocol run can be performed.

There are two more possible outcomes in addition to those two. The third situation is that the results are passing and the observer confirms that the cleaning process was done correctly. In that case, everyone is happy. The fourth situation is that all results are passing, but the independent observer notes that the operator did something wrong or different. Whoops! No one is smiling in this situation, because this is something like the situation where I get failing results and the operator did something incorrectly. There may be some cases where I might be able to say that the incorrect action by the operator did not affect the validity of the protocol. For example, suppose the SOP called for a 1% concentration of the cleaning agent, and only a 0.5% concentration was used. While I might accept that as a valid run, I would still have to do some retraining (and perhaps a change to the SOP to clarify how the proper concentration of the cleaning agent is prepared). There may be other situations where the incorrect action of the operator may have made the cleaning process more robust. For example, using the same example of the cleaning agent concentration, it might be that the operator prepared a 2% concentration instead of 1%. In that situation, the higher concentration might have been the cause of the acceptable residue results.

With all these possibilities, you might think it would be preferable to emphasize better cleaning process design *and* better training of manual cleaning operators, and just skip having an independent verification of the cleaning process. Certainly, better design and better training are *very desirable*. The purpose of the independent verification is to deal with situations where the cleaning protocol doesn't go as planned. It can be considered part of the "belt and suspenders" approach used in many aspects of pharmaceutical manufacturing. Certainly, that type of independent verification can be helpful in a *manual* cleaning protocol.

Some might object that "Of course, the operator will perform the cleaning SOP correctly if another person is closely watching the operator. But when the protocol is complete and cleaning is done routinely, who's to say the operator performs the SOP correctly". My answer to that possible objection is that as a validation specialist, I want to protect the integrity of the validation protocol, and having an independent verification can help with that. It gets back to what I said earlier, that the purpose of the validation protocol is to confirm that *if the cleaning process is done correctly, then I will get acceptable residue results*. If after completion of the protocol, the operator does not perform the cleaning process correctly, then the responsibility (or burden) for that falls with the production supervisor (or the production department). But that issue of correct performance of the cleaning SOP on a routine basis is something that is a possibility whether or not an independent observer is present during the validation protocol runs.

Some might be wondering whether an independent observer is needed for an automated cleaning process. If the cleaning process is a fully automated process like CIP cleaning, then it is *not* necessary to have an observer confirm that the operator selects the correct cycle and pushes the correct button(s). For an automated CIP skid, all those things, as well as information like times, temperatures, pressures, flow, and conductivity, are captured by the CIP skid itself. Therefore, if something goes wrong with the residue results, those operational parameters can be checked to confirm whether there was a problem with the cleaning process itself. If there are manual aspects to automated cleaning, there may be value to an independent observer confirmation. For example, in a parts washer, the loading configuration may be something to be independently observed. For things like transfer panels for CIP cleaning, if there is an automated confirmation of correct placement (or if incorrect placement is not possible), there may be little value in having an independent observer confirm correct placement.

Finally, the question sometimes comes up as to whether an independent observer using a manual cleaning worksheet should tell the operator immediately when the operator is doing something *incorrectly*. There are two views on this. One is to immediately tell the cleaning operator of the mistake so that it can be corrected without compromising the protocol, or so that the protocol run can be discontinued. The other is to let the operator continue, with the view that if a mistake was made at one step, let's try to see if the operator performs other steps incorrectly (trying to capture all the mistakes the first time rather than finding another mistake in a second protocol). Both views have some merit.

Let me be clear that a second-person verification of the correct performance of a *manual* cleaning process is *not* a regulatory requirement. However, most regulatory guidances emphasize the *variability* of manual cleaning processes and the need for more attention paid to such processes. A cleaning process worksheet used by an independent observer is one way to help address that extra attention.

The above chapter is based on a *Cleaning Memo* originally published in January 2019.

46 Dealing with Used and New Equipment

We generally know (at least we think we do) how to handle cleaning and cleaning validation for the manufacture of products we routinely make. We know what product is being cleaned and what potential next (also called subsequent or follow-on) products could be made in that cleaned equipment. So we can proceed to calculate carryover limits and then measure residues by appropriate sampling and analytical methods. On the other hand, what can or should be done for *used* equipment that I have purchased from another company or from a company specializing in used equipment? Or what can and should be done when I am purchasing *new* equipment? Remember my objective is to establish that the used (or new) equipment can be safely used for manufacture of my products.

We'll start with the situation of *used* equipment first. It is prudent (but not always possible) to obtain a history of products previously manufactured with that equipment. Questions include whether it is possible to identify specific products or product types that were processed. Were the products drug products? Were any of the products in the highly hazardous category (including beta lactams)? Were any non-drug products manufactured, such as cosmetics or agricultural products? That should be one of the first steps in terms of assessing whether the used equipment should be purchased, and may help direct steps to deal with potential contaminants left on the equipment.

If there are specific residues identified which should be evaluated, then various steps are possible. For example, if identified highly hazardous actives were processed, the equipment could be cleaned in a suitable cleaning process, and then those residues measured in a "cleaning verification" (one-time) study or protocol. Measuring those residues on the equipment *before* cleaning may also be done. The advantage of measuring residues before cleaning is that it tells us whether those actives of concern were actually present on the equipment at unacceptable levels. However, even if they were found acceptable before any cleaning, I would prefer to still clean and measure residues again *after* cleaning. The measurement techniques for highly hazardous active generally should be specific methods and not non-specific methods like TOC. The reason is that TOC will pick up any organic carbon sources, and generally will not provide a good (or accurate) picture of the presence of highly hazardous actives.

Regardless of the type of residues identified (or because of the lack of information on possible residues), I highly recommend an initial "pre-use" cleaning first using an alkaline detergent cleaning agent followed by a rinse and then cleaning with an acidic detergent cleaning agent followed by a more extensive rinse. The rationale

DOI: 10.1201/9781003366003-55

for the alkaline detergent cleaning is to remove soils which may be removed (by dissolution, emulsification, dispersion, suspension, or hydrolysis, for example) by a surfactant on the alkaline side. Once those soil types are removed, I then follow up with the acidic detergent cleaning agent to remove soils which may be removed by a surfactant on the acid side. If an *oxidant* may be of help, it is possible to use a peroxide additive to the alkaline detergent, or to follow up the acid detergent step with a peroxide or peracetic acid *polishing* step.

Following this pre-use cleaning, the equipment should be sampled for residues of the detergents used, as well as for any specific analytical tests based on identified actives of concern. Furthermore, it is expected that the equipment be visually clean. In addition, I generally recommend the use of TOC and conductivity for the final rinse to give me an overall picture of the cleanliness of the equipment, as well as possibly doing swab TOC measurements on expected worst-case locations. Finally, it may be possible to do non-specific testing by sampling with a suitable organic solvent and analyzing by FTIR or UV. What I would like to see is non-detectability by UV or a "flat line" by FTIR (insofar as a *flat* line is possible in FTIR) as a further confirmation of the effectiveness of the cleaning. Note that this latter technique may enable me to eliminate potential root causes if later I find "unknown peaks" in a subsequent study.

Once I have done this, my preference is then to clean the equipment with an already approved cleaning SOP in use in my facility, which hopefully will be the SOP I intend to use routinely on this "used equipment" going forward. This may also involve a "cleaning verification" to document the removal of the cleaning agent from the equipment.

Now let's turn to the situation with newly purchased "new" equipment (equipment that I am purchasing directly from the equipment fabricator). In this situation, I (hopefully) don't have to worry about drugs or other products made on the equipment. The major concerns are related to the fabricators processing materials (lubricants, cutting fluids, polishing agents, and the like) and to metal fines (which result from various fabricating steps).

Assuming that I don't have to worry about drug products (or other manufactured products) on the equipment, I would start with the alkaline/acid "pre-use" cleaning discussed for used equipment. The rationale for the use of both for new equipment is that the alkaline detergent will be more effective at removing organic processing materials and the acid product will be more effective at removing metal fines. Note that the use of the acid product first will not likely be as effective in that metal fines may be *physically trapped* in organic residues and may not be adequately removed by the acid detergent.

I would still recommend the non-specific analysis by TOC and conductivity, as well as the visual assessment and the evaluation by FTIR or UV to document the effectiveness of the overall cleaning process. Once I have established the lack of unacceptable residues on the new equipment, I would do the same cleaning with the expected routine cleaning SOP to be used on the equipment (as suggested for "used" equipment).

The suggested relatively elaborate cleaning evaluation may be considered over-kill in many situations. However, the objective of this one-time cleaning is to get it right the first time. This avoids situations either where I have to reclean and retest, or where I might have unforeseen problems come up in the future (potentially disrupting routine manufacture or potentially calling into question manufactured product quality).

The cleaning and evaluation process should be appropriately documented, both for regulatory inspection reasons as well as to provide a documented history of the equipment.

Note that in both situations (used equipment and new equipment), there are other activities that should occur (as part of IQ/OQ, for example) before I turn the equipment over for routine production use.

Furthermore, the examples discussed are generally relevant for stainless steel equipment cleaned with aqueous solutions. For other situations, there may have to be some modifications to the suggested approach, but the principles behind the approach should be more or less the same.

The above chapter is based on a *Cleaning Memo* originally published in April 2018.

47 Solving Cleaning Validation Problems by Analogy

I have been very interested in recent popular discussions of "models", specifically models *in scientific thinking*. We, in the cleaning validation community, have a certain model of what happens in a cleaning process that could lead to cross-contamination of the next manufactured product. That model usually envisions that residues left on the cleaned equipment surfaces will transfer more or less *completely* and *uniformly* to the next manufactured product. How do we know this? Well, it seems like a reasonable scenario, even though we know there are probably situations where it is not exactly true. But, it is the model we use in designing our cleaning validation program. Now, you are probably asking yourself "What does this have to do with solving problems by analogy?"

Okay, thinking of "models" caused me to think of one of my favorite books dealing with the nature of science, *The Structure of Scientific Revolutions*, by Thomas S. Kuhn. Kuhn, who is probably the premier historian of science, wrote this book in 1962. When I was an undergraduate at Michigan, we had a short class (one credit, I believe) that covered this book and Kuhn's model of how science works, and specifically how scientific "revolutions" happened. So, I recently took the opportunity to reread the book (actually a second enlarged edition of 1970) [Kuhn, 1970], and it was truly refreshing. Kuhn talks about how scientists ordinarily work under a paradigm (think "model") of how the world operates (at least within their own scientific discipline). And, because a paradigm always has "loose ends", the primary scientific endeavor within a given paradigm is puzzle solving, or in essence addressing all the loose ends that a paradigm entails. Kuhn addresses the issue of trying to solve a problem for which the scientist has no literal equivalent before. How is this done? Here is Kuhn's answer:

> A phenomenon familiar to both students of science and historians of science provides a clue. The former regularly report that they have read through a chapter of their text, understood it perfectly, but nonetheless had difficulty solving a number of problems at the chapter's end. Ordinarily, also, those difficulties dissolve in the same way. The student discovers, with or without the assistance of his instructor, a way to see his problem as *like* a problem he has already seen. Having seen the resemblance, grasped the analogy, between two or distinct problems, he can interrelate symbols and attach them to the nature in the ways that have proved effective before.

DOI: 10.1201/9781003366003-56

Kuhn goes on to say "Scientists solve puzzles by modeling them on previous puzzle solutions…" The examples he discusses are related to mathematical equations used in the physical sciences. But, the principle of looking for analogies is helpful in non-mathematical puzzling problems. So, when you come across a puzzle (problem) in cleaning validation, think of things that might be similar, and see if you can come to an analogous solution.

Here is an example. If you follow my writings on swab sampling recovery studies, you are probably aware that I teach (or I believe) that as the amount of residue on the surface increases, then (other things being equal) the recovery percentage decreases. How do I know that? Well, I don't have a lab, so I have not done any clear experiments to demonstrate that (although I have proposed experiments to do so). One possible analogy that might be appropriate to support my contention is the actual cleaning of soils (think of your product as a soil, even though most don't like to call it that) on equipment surfaces. What is easier to clean, a surface with a large amount of soil, or a surface with a much smaller amount of soil? I think we all might intuitively say "the surface with the smaller amount" (and that is also based on prior experience). That is, other things being equal, the same cleaning process might remove "all" of the soil in the situation with the smaller amount of soil, but leave some soil behind in the situation with the greater amount of soil. So, in that latter situation, I would have to use a more aggressive cleaning process.

How does this apply to swab sampling recovery? Isn't a cleaning process completely different from a swab sampling process; after all, the objectives are different? Yes, they are different, but there *are* similarities. In both a cleaning process and a swab sampling process, I am essentially (here is *my* analogy) removing a material from a surface. In one case, I am removing it in the cleaning process so the material can be safely disposed of; in the other case, I am removing it in a swabbing process so the material can be analyzed by my HPLC (or TOC) process. If the larger amount of soil on the surface makes it more difficult to remove a larger percentage of that soil in a cleaning process, might it not also be the case that the same principle holds true for swab sampling? In other words, even though the purpose of a cleaning process and a swabbing process are fundamentally different, from a process perspective, there are similarities. (Note that this similarity is something I also point out in my webinars as to the variability of swabbing because swabbing is like a type of manual cleaning and has elements of variability present in manual cleaning procedures.)

Are there other situations in cleaning validation where such analogies may be useful? I think so! Think about a Clean Hold Time (CHT) and look for a similar situation which might suggest that a formal CHT protocol is not required if the equipment is stored dry and protected from external sources of bioburden. Thinking about a situation and doing an appropriate risk assessment is often a better approach to cleaning validation "puzzles" as compared to a "cookbook" answer.

The above chapter is based on a *Cleaning Memo* originally published in July 2020.

REFERENCES

Kuhn, T.S. *The Structure of Scientific Revolutions*, 2nd edition, University of Chicago Press, Chicago, IL, 1970.

48 Causing Cleaning Validation Problems by Analogy

Chapter 47 covered how analogies can be useful when we are faced with a new cleaning validation issue. In this chapter, we will look at how analogies can do just the *opposite* – lead us down a rabbit hole that just may seem to be a solution, but in reality, just complicates life. A key issue in using analogies is that there has to be a reasonable *similarity* between a situation with a well-established approach and the new situation that we are trying to address. That is, we can say that "B looks like A; therefore what works for A will also work for B"; but in reality, while B "*looks like*" A, in reality, it is *different enough* such that what works for A probably is not applicable to B. Here are three situations that we may have all come across that illustrate this lack of similarity.

The first involves analytical method validation for the analysis of residue samples in solution (which might be either a solution in an extracted swab sample or a rinse sample). How do we address what to establish as a *linear range*? Well, we might look at what is done to establish a range for the active concentration in a *potency* assay for a drug product (by potency assay I mean an assay for the concentration of the active in the drug product). In that validation for potency, the analytical lab knows what the target concentration is and then validates the range below and above that value. That range may be 75–125% of the target value, or 50–150% of the target value. Okay, that is well accepted for a potency assay [ICH, 2005]. So, now I am faced with analytical method validation for residue measurement in a cleaning validation protocol. So, the two situations (analytical method for potency assay and analytical method for residue determination) are similar (right?). Therefore, for my analytical method validation for cleaning validation purposes, I merely validate a linear range from 50% up to 150% *of the residue limit* in the solution. Sounds easy enough, and it is something that I see much too often.

However, the issue is that the two situations are *not* exactly similar enough. In the case of the potency assay, the *target* is the amount I am hoping to find (the 100% value). In the case of residue assay, the *limit is not my target* (or if it is my "target", it is a target I am hoping to miss!). The limit in a residue analysis is a value I want to be *below*, and hopefully *significantly* below. If the bottom of my linear range is 50% of the limit, and if that value is actually my LOQ (limit of quantitation), then I am really in a situation where, if I want a robust cleaning validation process, the best I can say (assuming 100% recovery in sampling and a

DOI: 10.1201/9781003366003-57

residue value below the LOQ) is that the residue is less than 50% of the limit. While technically that works from a compliance point of view, most companies want to demonstrate a more robust cleaning process.

A second difference is that for the residue assay, my "target" value is not a fixed value, but may be a *range*. What do I mean by that? What I mean is that my desired values (to demonstrate the robustness of my cleaning process) might be values about 10–30% of the residue limit. So my linear range for the residue in solution might be 10–100% of my residue limit, which is a much wider range than for a potency assay. You might wonder why I extend the upper limit up to 100% of the limit (particularly if I prefer that value be much lower). The answer is that while I prefer lower values, I don't want to exclude the possibility that in a few cases I will get higher values. Unless the analytical method is validated to measure at 95% of the limit (still a passing value), how can I trust the validity of that value that is 95% of the limit? By having the method validated up to 100% of the limit, I am covering the possibility that I might have values in that range (even though those higher values are *not* preferred).

So the "take home" less here is that for analytical method validation, the *target* value in a potency assay is different enough from a *limit* value in a cleaning validation residue assay such that the *range* for my method validation should probably be different. Note that this is not to say that *other* aspects of analytical method validation for potency assays might not be applicable.

A second situation of where an analogy comes up short is that of doing swab recovery studies at different spiked levels. That is, since we do analytical method validation over a range of values for residues in solution, we should also do the recovery studies over a range of values (such as 50–150% of the residue limit). The main issue I want to address here is whether there is sufficient similarity between the purpose of a range that is used for the analytical methods in solution and the purpose of a range that is typically used for swab recovery studies. For the residue values in solution, the purpose typically is to establish a linear range where measured values are accurate and precise. With residue recovery studies, are we expected to find a linear range? Perhaps, but the data I have seen sometimes shows that the recovery percentage is the same over the range, sometimes shows recovery increasing with increasing spiked levels, sometimes shows recovery decreasing with increased spiked levels, and sometimes shows a highly variable relationship [LeBlanc, 2013]. Now, it is possible that all of these may be true, and that the relationship of percent recovery to spiked level is highly dependent on the residue, the surface, the analytical method, and the swabbing technique. And while I acknowledge that the recovery percentage may vary based on those factors, I believe (based on reason and logic) that in general, *over a relatively narrow range*, the recovery percentage will be the same in a carefully controlled experimental design. (See Chapter 29 of this volume.) That is, I don't expect a significant variation of recovery percentage over narrow spiked levels of 1X to 5X. However, if the range increases from 1X to 25X, I believe as the residue load increases on the surface, the percentage recovered in swab sampling decreases (even though the amount removed would increase). Therefore, performing a recovery study only at the limit should represent a worst case (applicable to all values up to that limit).

A third example of where analogies can go wrong involves the statistical analysis of swab sample results. Here is the analogy. Cleaning validation is like process validation. In process validation, I look at the results of a certain quality parameter of the manufactured product, and statistically analyze the results based on samples *within* a batch and samples *from batch to batch* to determine consistency [FDA, 2011]. So, if cleaning validation is similar to process validation, I can then statistically evaluate the various swab samples to determine the consistency of my cleaning process. The issue with this analogy is that consistency among analytical values for product quality consistency (I am expecting only minor variation) in process validation is different from the consistency expected of swab samples in cleaning validation. My expectation for consistency in cleaning validation is that "swab samples results are *consistently below the calculated limit*". Swab sampling locations are not the same "population" (some are harder to clean locations and some are easier to clean locations), therefore treating these different locations by statistical analysis doesn't make sense. If I really wanted a higher level of consistency, the best I could hope for is that all my results were non-detectable (below the limit of detection). Even in that situation (all non-detectable), I certainly could not apply statistical analysis. Treating different swab locations in a statistical analysis has the aura of a "scientific" approach, but in most cases, it is just "window dressing".

So, where does this leave us in the application of analogies? Clearly, there has to be an element of thought, analysis, and wisdom in deciding which analogies are useful and which are not. This is not unlike other situations we face in everyday life.

The above chapter is based on a *Cleaning Memo* originally published in August 2020.

REFERENCES

ICH Q2(R1), "Validation of Analytical Procedures: Text and Methodology". November 2005.

LeBlanc, D.A., "Swab Sampling Recovery as a Function of Residue Level", in *Cleaning Validation: Practical Compliance Solutions for Pharmaceutical Manufacturing, Volume 3*. Parenteral Drug Association, Bethesda, MD, 2013, pp. 181–185.

U.S. Food and Drug Administration, *Process Validation: General Principles and Practices (Revision 1)*, U.S. Government Publishing Office, Washington, D.C., January 2011.

Appendix A
Acronyms Used in This Volume

ADE	Acceptable Daily Exposure
ADI	Acceptable Daily Intake
API	Active Pharmaceutical Ingredient
APIC	Active Pharmaceuticals Ingredient Committee
ARB	Angiotensin Receptor Blocker
BS	Batch Size
CAPA	Corrective And Preventive Action
CFU	Colony Forming Unit
CGMP	Current Good Manufacturing Practice
CHT	Clean Hold Time
CIP	Clean In Place
CMO	Contract Manufacturing Organization
CV	Cleaning Validation
CVL	Cleaning Validation Limit
DHT	Dirty Hold Time
EMA	European Medicines Agency
ER	Extended Release
FDA	Food and Drug Administration
FTIR	Fourier Transform Infrared
GMP	Good Manufacturing Practice
HBEL	Health-Based Exposure Limit
HPLC	High-Performance Liquid Chromatography
ICH	International Conference for Harmonization
IMP	Investigational Medicinal Product
IPA	Isopropyl Alcohol
IQ	Installation Qualification
ISPE	International Society for Pharmaceutical Engineering
LD50	Lethal Dose Fifty Percent
LOD	Limit Of Detection
LOQ	Limit Of Quantitation
MAC	Maximum Allowable Carryover
MACO	Maximum Allowable Carryover
MDD	Maximum Daily Dose
MOC	Material Of Construction
MSA	Methane Sulfonic Acid
NDEA	Nitrosodiethylamine
NDMA	Nitrosodimethylamine
OEB	Occupational Exposure Band

OEL	Occupational Exposure Level
OOS	Out Of Specification
OQ	Operational Qualification
ORA	Office of Regulatory Affairs
PDE	Permitted Daily Exposure
PIC/S	Pharmaceutical Inspection Cooperation Scheme
PPE	Personal Protective Equipment
PPQ	Process Performance Qualification
PTFE	Polytetrafluoroethylene
PV	Process Validation
QA	Quality Assurance
QRM	Quality Risk Management
SA	Surface Area
SDA	Safe Daily Amount
SDS-PAGE	Sodium Dodecyl Sulfate – Polyacrylamide Gel Electrophoresis
SEA	Swab Extraction Amount
SSA	Shared Surface Area
SSR	Separate Sampling Rinse
SIP	Steam In Place (or Sterilize In Place)
SOP	Standard Operating Procedure
TAMC	Total Aerobic Microbial Count
TCYMC	Total Combined Yeast and Mold Count
TFF	Tangential Flow Filtration
TOC	Total Organic Carbon
TTC	Threshold of Toxicological Concern
USP	United States Pharmacopeia
UV	Ultraviolet
VL	Visual Limit (or Visual Level)
VRL	Visual Residue Limit (or Visual Residue Level)
WFI	Water For Injection

Appendix B
Shorthand Notations for Expressing Limits

In this volume, I utilize five categories for expressing limits. All of these are useful in discussing limits for different purposes. Following are the shorthand "abbreviations" I use.

L0: The safe amount of residue that can be administered to a person on a daily basis for a long time period. The units for this are mass units such as µg or mg. Note that L0 values for a given residue may be different depending on the route of administration (e.g., oral, injectable, topical). Furthermore, L0 may be adjusted based on a limited time of potential exposure. L0 may be calculated by formulas such as 0.001 of minimum daily dose of an active, by the Risk-MaPP Acceptable Daily Exposure (ADE) determination, by the EMA Permitted Daily Exposure (PDE) determination, or by an LD_{50} determination (for cleaning agents).

L1: This is the safe concentration of residue in the next manufactured product. This is typically in units such as ppm, µg/g, or µg/mL. This, of course, makes certain assumptions about the route of administration and time frame of exposure in determining the L0 amount. For finished drug product manufacture, it is calculated by dividing the L0 value by the maximum daily dose of the next drug product. Note that for those who use a "default" value of 10 ppm in your carryover calculations, that 10 ppm value is typically used as the L1 value whenever it is *more stringent* than the calculated L1 value.

L2: This is the total amount of residue allowed in a batch of the next product, and therefore the total amount allowed on shared product contact surfaces of the equipment train. This is typically in mass units such as µg, mg, or g. It is calculated by multiplying the L1 value by the minimum batch size of the next manufactured product.

L3: This is the amount of residue allowed per surface area of product contact surfaces. This is typically in units of mass per surface area, such as $µg/cm^2$ or $µg/in^2$. It is calculated by dividing L2 by the total product contact surface area of the equipment train.

L4: There are several variations here, depending on how sampling is done.

L4a is the amount of residue per swab (for swab sampling). It is in mass units, typically µg or mg. L4a is calculated by multiplying the L3 value by the area swabbed (typically 25 cm^2 or 100 cm^2). For clarification, if two swabs (for example, one wet and one dry) are used, L4a is not the amount per swab, but the amount for two swabs. [Note that the L4a limit can be increased by swabbing a larger surface area.]

L4b is the concentration limit in the liquid (such as water or solvent) the swab is extracted into. This is typically in units such as ppm, μg/g, or μg/mL. It is calculated by dividing the L4a limit by the amount of liquid used to extract the swab. [Note that the L4b limit can be increased by extracting the swab in a smaller volume of liquid.]

L4c is the concentration limit in the liquid (such as water or solvent) that is used for rinse sampling. This is typically in units such as ppm, μg/g, or μg/mL. If the entire equipment train is rinsed as a unit, L4c can be calculated by dividing L2 by the amount (or volume) of liquid used to sample the equipment in the rinse sampling process. If items in the equipment train are sampled independently of each other, then L4c is calculated by multiplying L3 by the surface area of the equipment sampled and then dividing that result by the amount (or volume) of liquid used to sample the equipment sampled in the rinse sampling process. [Note that the L4c limit can be increased by performing the rinse sampling with a smaller volume of liquid.]

Note that other terms can and have been used in the industry. However, the terms L0, L1, L2, L3, and L4 have no significant connotations (except that they are associated with me), so that is one reason I prefer them. A second reason for using them is that terms like MAC (originally Maximum Allowable Carryover, referring to the total amount of residue that can be carried over from the cleaned equipment to the next manufactured product) have been used in so many different ways by different companies that it seems futile to try to have people in the industry use the term in the same way (incidentally, MAC as originally defined is the same as my L2 value).

One reason I utilize these *different* shorthand expressions is that which limit is useful will depend on what I am doing. For example, if I am doing sampling recovery studies, I find it useful to refer to the L3 limit, because this limit is what I want to use for my one spiking level in my recovery study. Furthermore, if I am doing a product grouping (matrixing), I want to perform a protocol with the most difficult to clean product evaluated at the lowest limit of any product in the group. A comparison of L3 values will most easily determine the lowest limit in a product group. As mentioned, if I want to use a 10 ppm default value when it is more stringent than my carryover calculation, I compare the 10 ppm to my L1 value (and not my L4b value). If I am using "stratified sampling", then my L2 limit is critical. Note that in some cases, alternatives can be easily used; companies using swab sampling can express that limit as either L4a or L4b; which is chosen is usually a matter of past practice.

Index

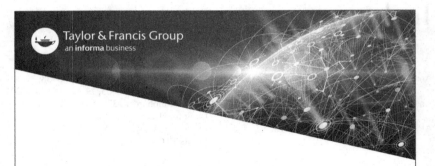

Printed in the United States
by Baker & Taylor Publisher Services